AN AGRICULTURAL ATLAS OF SCOTLAND

An Agricultural Atlas of Scotland

J. T. COPPOCK

Professor of Geography
University of Edinburgh

JOHN DONALD PUBLISHERS LTD
EDINBURGH

© J. T. Coppock, 1976
Published by
John Donald Publishers Ltd.,
8 Mayfield Terrace,
Edinburgh EH9 1SA, Scotland

ISBN 0 85976 016 2

Printed and bound in Great Britain by
Morrison & Gibb Ltd., London and Edinburgh

Preface

IN 1965, when I was appointed to the Ogilvie Chair of Geography in the University of Edinburgh, I decided that an up-to-date agricultural atlas of Scotland was needed. Nearly forty years had elapsed since H. T. Wood had prepared the first and, so far, only agricultural atlas of the country and much had changed in the interval; moreover, his atlas had long been out of print. I envisaged the atlas as a companion volume to my *Agricultural Atlas of England and Wales*. The Leverhulme and Carnegie Trusts generously gave financial support and a research assistant (Mr J. McG Hotson) and a trainee draftsman (Mr M. Fordham) were appointed. At this stage, the systems of automated cartography used in the preparation of the second edition of my atlas of England and Wales were still in an experimental stage and it was decided to produce hand-drawn maps, though computer print-outs were used to speed the process of map compilation. The year 1965, the latest for which data were then available, was chosen as the base year and work began effectively in 1967. By 1969, most of the maps were drawn and I confidently predicted, in an article in the *Cartographic Journal* of June 1969, that the atlas would soon appear in print. However, at the suggestion of a prospective publisher (who did not, in fact, produce the book) it was decided to give the atlas an historical flavour by comparing the situation in the mid-1960s with that in 1870 and in 1938, and publication was accordingly delayed while the relevant data were abstracted, the maps compiled and the text written. Further delays resulted from my appointment in 1971 as a specialist adviser to the Select Committee on Scottish Affairs for its investigation into land resource use in Scotland and from inflation and the crisis facing the printing and publishing industries. While my optimism in 1969 has proved unfounded, I believe that the provision of a historical perspective has been justified.

It was my original intention to have an extended and continuous commentary on the maps, similar to that in my atlas of England and Wales; but, as the atlas progressed, I felt that there would be merit in always having the maps and appropriate text on the same or facing pages. In practice, it has not been easy to match text and the space available, for the literature is very uneven, and the layout often leaves no alternative between writing half a page and writing two and a half pages. On some occasions I have provided only some basic statistics and an indication of the main features of the map. On others, where the topic is both well documented and interesting, as with beef cattle and sheep, I have had to omit material which I should like to have included. Fortunately, it is the function of an atlas to point to further research as well as to suggest answers to present questions, though the questions which can be posed are restricted by the nature of the agricultural census. Because of this layout, with the text very largely on the same pages as the illustrations and tables to which it refers and complete listings of the 283 maps and 89 tables, I have not thought it necessary to provide an index.

The atlas is intended for a wide audience and I have therefore avoided technical terms as far as possible; I would, however, ask readers not to accept maps uncritically at their face value but to consider the nature and limitations of the material being mapped. I would also ask those who believe that the maps record the obvious, what mental image they had of the distribution before they saw the map. I hope the atlas will be of interest to administrators, farmers and politicians, to those in various branches of education, and to the members of the public at large, many of whom (if a sample of students is any criterion) have little knowledge of what goes on in the 98 per cent of Scotland which is officially countryside. One day I hope to produce my ideal atlas, where maps are up-to-date and readily updated, where not only distributions at particular points in time are mapped but also changes over various periods of time—seasonal, annual and decennial, and where I can escape from the straitjacket of the parish. In the meanwhile, I hope that this atlas will serve to fill a gap of which I have been only too well aware in my own reading and research.

J. T. COPPOCK

Acknowledgments

THIS atlas has been made possible with the financial support of the Carnegie, Leverhulme, MacRobert and Frederick Soddy Trusts and of the Scottish Landowners Federation, and this I gratefully acknowledge. Equally important has been the range of assistance given by the University of Edinburgh, notably in the provision of facilities for computing. Many individuals, too, have contributed to the production of this atlas and I am grateful for their help. Pride of place must go to Mr J. McG Hotson, MA, who was appointed as a research assistant in 1967 and has helped in various ways ever since, though his obligation to do so has long ceased; he devised the computer programs used for map compilation, plotted many of the compilations for the dot maps and the maps of livestock movements, and made many helpful suggestions. I am grateful, too, to Mr Carson Clark, formerly Chief Technician in the Geography Department, and Messrs Alec Bradley, Michael Fordham and Ray Harris who drew the final maps; to Miss Susan Bonus, Mr Richard Gillanders, Mrs Barbara Morris, Miss Elizabeth Thompson and Dr Lillian Walker, who helped with the compilation and checking; and to Mrs S. Allen, Mrs V. Reid, Mrs P. Paterson and Mrs P. Robertson, who typed the manuscript. The raw material has been derived mainly from the annual agricultural censuses, and I am particularly grateful to Mr O. J. Beilby, formerly Chief Economist in the Department of Agriculture and Fisheries for Scotland, to Mr P. M. Scola, formerly head of the Statistics Division, to Dr J. M. Dunn, their successor, and to Mr D. R. Dickson of that division, for their help; Mr Scola, in particular, gave most valuable advice on the composition of the atlas (though he bears no responsibility for its final form). Mr J. P. Struthers, then Senior Inspector in the Department of Agriculture, and his colleagues in the inspectorate kindly provided information on movements of livestock and on seasonality of farm activities, for which no other data exist. Other information was provided by Dr J. D. W. McQueen, Marketing Officer of the Scottish Milk Marketing Board, Dr R. W. Gloyne of the Meteorological Office, Dr R Glentworth, Head of the Soil Survey in Scotland, Dr M. Pyke of the Distillers Company, and many others, to all of whom my grateful thanks are due; Dr W. J. Carlyle, of the University of Winnipeg, generously allowed me to use material incorporated in his pioneering PhD thesis on the geography of livestock movements in Scotland. As usual, my wife has exercised her critical skills in correcting the manuscript, suggesting many improvements to the text and reading the proofs. Lastly, I am grateful to Messrs John Donald Ltd, especially Mr John Tuckwell, for producing the atlas so quickly once they had agreed to publish it. I hope that all who have helped will feel some sense of satisfaction that this project has at last been completed, though any errors, whether of fact, computation or interpretation, are mine alone.

J.T.C.

List of Figures

List of Tables

Contents

Introduction: Sources and Methods

THIS atlas is based very largely on parish summaries of the returns from the main agricultural censuses taken in June 1870, June 1938 and June 1965, and it is therefore important to appreciate both the nature and the limitations of data from these sources. In this discussion, attention will be focused primarily on present characteristics of such data, features peculiar to 1870 and 1938 being discussed in chapter 7, which deals with historical change.

Agricultural returns in 1965 were obtained from all occupiers of holdings of more than one acre of agricultural land, a lower limit that has obtained since 1891; in 1870 the minimum acreage was $\frac{1}{4}$ acre. Occupiers of such land are required by law to make a return when asked to do so by the Department of Agriculture and Fisheries for Scotland, but from 1866 to 1918 and from 1921 to 1925 the returns were voluntary. A great deal therefore depends on the interpretation of 'agricultural' and on the adequacy of the Department's records of holdings. The latter are now believed to be reasonably complete, but not only does the distinction between agricultural and non-agricultural represent merely a point in a continuum, but administrative practice has changed considerably in the 110 years since the regular collection of agricultural statistics began in 1866—a consideration that must be kept clearly in mind when attempts are being made to compare changes over time and to monitor trends. 'Agricultural' in a census context includes not only land under crops and grass and rough grazings, but also woodlands that are part of farms, farm roads, steadings and other land uses necessary for the functioning of each farm; since 1959 it has also included all deer forests on farms whether such land is grazed or not.

The returns for 1965 relate to the conditions on each holding on 4 June in that year. Occupiers were asked to record the acreages under different land uses and the numbers of different classes of livestock on their holdings on that particular day. After checking, the results of the individual returns in each of the 891 parishes are totalled to provide parish summaries and these in turn are added together to give figures for each county and region and for Scotland as a whole. The parish summaries are thus not prepared as ends in themselves, but are merely an administrative convenience and an historical legacy from the beginnings of the census; for such summaries have been prepared annually since 1866, and although they do now have some value for analytical work, this is not their main purpose. The individual returns are confidential, but the summaries, which are deposited in the Scottish Record Office in Edinburgh, are available for consultation by anyone who wishes to do so; only the county, regional and national totals are published, in the annual volumes of agricultural statistics.

THE AGRICULTURAL CENSUS

The census form for June 1965 contains entries relating to 125 items: of these, 16 relate to crops, 10 to grass, 4 to other land uses, 23 to horticultural crops, 42 to livestock and 9 to labour. The crops are those in the ground on census day or those for which the land has already been prepared; some vegetables are planted later and so escape enumeration in the census, although this is not an important problem in interpreting the Scottish agricultural census. The acreage under orchards, however, is not recorded in June, but is enumerated in the December census, although this, too, is a minor feature of agricultural land use in Scotland. Occupiers of agricultural land are requested to record acreages correct to the nearest $\frac{1}{4}$ acre and to give the full acreage of the fields, viz that recorded on the large-scale plans of the Ordnance Survey, including headlands and those parts of surrounding hedges, banks and walls which can properly be attributed on this basis to each field.

The labour force and the livestock are also those on the holding at the time of census, though occupiers are instructed that, where livestock are temporarily away from the holding, they should be returned under that holding. A date in early June is less satisfactory for recording numbers of livestock than for recording crop acreages because of the mobility of livestock and the large seasonal movements that take place both locally and over long distances; thus, on the basis of the June census, beef cattle appear to have a less important place in the farm economy of east central Scotland than an assessment of average numbers throughout the year would indicate, largely because of the importance of winter fattening in that part of the country. The relationship between 4 June and the seasonal cycle of livestock production is also a factor. It seems likely that the number of lambs on some hill farms is underestimated because farmers have not completed counting their flocks by the census date.

The kinds of information contained in the census are largely a reflection of its long history and of the fact that it is conducted by means of a postal questionnaire, which is completed unaided by the individual farmer. Apart from the fact that the census forms have become progressively more complicated, the character of the census has not changed radically since it was first undertaken in 1866; in general, data have not been requested which assume that farmers keep records, eg of the numbers of fat lambs sold off the holding during the year. Confining data to the situation on a given day may facilitate collection, but there are consequent problems of interpretation.

The usefulness of the census as a source for the compilation of an agricultural atlas of Scotland depends on a number of characteristics of the data: in particular, the accuracy and reliability of the data, the extent to which the information can be interpreted in agriculturally significant terms and the spatial characteristics of the data. No assessments of reliability are

published, but the Department undertakes plausibility tests to check the consistency of the information provided by farmers. It is probable that the most accurate information is that relating to acreages of crops and grass, for these are grown in relatively small enclosures, the areas of which can readily be obtained from the large-scale plans of the Ordnance Survey and are likely to be shown on leases or deeds. It is true that recording the total acreage of each field, as given on the Ordnance Survey plans, exaggerates the area actually under a given crop, while the planimetric view leads to underestimation on sloping ground; but this over- or underestimation does not affect to any marked degree the relative importance of the different crops. It is possible that the acreages of permanent grass are less accurately recorded than those of crops, but in view of the importance of temporary grass in Scotland, any discrepancies are unlikely to be very important; in any case, the distinction formerly made between permanent and temporary grass has now been abandoned in favour of a division on the basis of age. The acreage under rough grazing is undoubtedly the least accurately recorded, for it is not itself of any great significance to the individual farmer, who is more interested in the stock which graze it; furthermore, he is often ignorant of its true extent, since much of the land under rough grazing is not covered by the Ordnance Survey's large-scale plans on which the size of individual parcels is recorded. It seems likely that many of the figures quoted are residuals, obtained by deducting the known acreages of the more intensive uses from the total acreage of the farm, and some of the land thus recorded as rough grazing may not in fact be grazed at all or used for any agricultural purpose. The existence of large acreages in crofting tenure in the seven crofting counties (Argyll, Caithness, Inverness, Orkney, Ross and Cromarty, Sutherland and Zetland), especially the large tracts of common grazing, is an added complication. It is also impossible, in the absence of cadastral records such as exist in most European countries, to establish whether any land is omitted altogether because it is claimed by no farmer or because its occupier has not been asked to make a return; J. Fraser Hart reported one instance where the ownership of $1\frac{1}{4}$ square miles of hill land was unknown and was not claimed by any of the farmers whose territory surrounded it.

The accuracy with which numbers of livestock are recorded must depend on farmers' own estimates and records, for there is no independent source, such as farm leases or the Ordnance Survey's large-scale plans, to which the farmer can turn for confirmation. The introduction of subsidies, notably the Hill Sheep and Hill Cattle Subsidies, is likely to have led to improvements in the accuracy with which livestock numbers are enumerated by providing an incentive to keep accurate records, but no such check exists in respect of the smaller livestock, which are probably the least accurately recorded of all the livestock. In part, of course, the accuracy of these data on livestock (as with those on crops) receives confirmation or otherwise from the plausibility of the picture they provide, although such confirmation is not conclusive.

Interpretation of the acreages and numbers provided by the census depends both upon the consistency of the interpretations made by farmers throughout the country and also on the agricultural significance of the data recorded. The interpretation of rough grazing, for example, is likely to vary somewhat throughout the country, depending on the relative importance of improved and unimproved land. The results of the Scottish census are probably more readily interpreted than those from the similar census of England and Wales because of the greater importance of grazing livestock, especially beef cattle and sheep, and because the whole cattle population in Scotland is divided into beef and dairy cattle, whereas only cows and heifers are so divided in England and Wales. Nevertheless, in interpreting both censuses, assumptions have to be made about the purposes for which different classes of livestock are kept and different crops grown; for example, barley cannot be subdivided into that grown primarily for sale and that consumed on the farm or between that grown for stock-feed and that grown for malting and other purposes. Difficulties of this kind arise even when the records of individual holdings are available, so that assumptions can be checked against the features of the farm economy. Where these records have already been summarised by parishes, additional problems occur because it is impossible to obtain such confirmatory evidence or to tell how far different enterprises are common to large numbers of holdings or occur on quite different types of holdings within the same parish. In these circumstances it is possible to consider the characteristics only of the 'parish farm', while recognising that this may be a compound of quite different types.

MAPPING THE AGRICULTURAL CENSUS

Probably the most important attributes of the census data for the production of an agricultural atlas are their spatial characteristics, in respect of both the holdings themselves and the parishes by which they are summarised. Relatively little is known, at least publicly, about the layout of individual holdings, for the census record is related only to a farm address, which may consist of no more than the name of the farm and that of the parish in which it lies. The Department of Agriculture does have maps of farm boundaries which were prepared in the 1940s, but they are not kept up to date and are not in any case related to the census data. There is thus no indication from the census whether the land in a particular holding overlaps the boundaries of the civil parish by which the records of holdings are summarised. There is, of course, no necessary relationship between parish and farm boundaries and any discordances between the two are likely to be increasing as farms are enlarged by amalgamation or by the purchase or leasing of additional land. It is highly probable, at least in the smaller parishes, that the farmland enclosed by the parish boundary does not correspond exactly with the land in those holdings which provide the data in the parish summary. Some land in a given parish is likely to form parts of farms which are returned under other parishes, often those in which the bulk of the farm territory lies; similarly, parts of the farms included in a given parish summary are likely to be located in other, probably adjacent parishes. The trend towards the creation of linked or led farms, ie those under the same management, but often physically separate, is likely to have made this a more significant consideration in recent years; for while such farms will often comprise hill and low ground farm in the same parish, parts of other led farms may be located in different parishes and even in different counties. It was estimated in 1972 that about a quarter of full-time holdings in Scotland were linked in this way, and the topic is under active investigation by staff of the Department of Agriculture and Fisheries for Scotland. How important the trend is for the interpretation and mapping of census data cannot be established; administrative practice in respect of the return of amalgamated holdings seems to have varied, and while the general trend has been to encourage farmers to make a single return where holdings are farmed together, the decision whether such holdings are in fact separate or parts of multiple holdings (of which led farms are a special case) is one for the farmer alone. The importance of these problems must not be exaggerated, for it seems likely that the great majority of land returned under all but the smaller parishes refers to land lying within the boundaries of the parish in question, an assertion that also receives some support from the plausibility of the resulting maps. Nevertheless, it must also be recognised there is an element of uncertainty in the census records which cannot be resolved from the available evidence, and it may well be that some of the apparent anomalies that appear on the individual maps in the following pages could be explained by such discordance between the location of the land to which the summary data refer and that of the civil parish.

Such discordance, if it exists (and the occurrence of parishes where the agricultural area is greater than that of the civil parish indicates that it does), is likely to affect chiefly the detail of the maps. Much more significant for their cartographic representation is the character of the parishes themselves. The 891 parishes in Scotland vary greatly in size, shape and physical homogeneity. The smallest parish is less than 100 acres (40ha) and the largest over 250,000 acres (100,000ha), so that the range of generalisation represented by the parish summaries is very large. It is true that much of the land in these very large parishes is rough grazing and that the component farms themselves are frequently very large (measured by their area), so that the differences in the degree to which holding data

are generalised are less than would at first appear from a comparison of their absolute size; but even if the comparison is restricted to the acreage under crops and grass, the range is still very large, from 0 to nearly 25,000 acres (10,000ha), while the number of holdings per parish ranges from 4 to 1,354.

Perhaps even more important for their cartographic representation than mere size is the relationship of parish territory to the physical environment. As a general rule, the larger the parish, the greater the range of soils and climates it is likely to contain. Most of the larger parishes in Scotland contain both upland and lowland, and the disadvantages of such heterogeneity are particularly acute where parishes are long and narrow, as with Urray; there are several instances in the succeeding maps where patterns of distribution are distorted by apparently anomalous values for such long parishes, notably in north Scotland. The general effect of such physical variety within parishes is to blur physical contrasts that are readily observable on the ground, though it must also be remembered that the individual farms themselves may have a similar shape or incorporate a similar range of land-types.

The characteristics of the parishes and of the parish summaries have several implications for mapping the data and for interpreting the results. First, the actual boundaries of the parishes are likely to have no particular significance and emphasising them may well contribute to a misleading impression of the actual distribution being mapped; for this reason, and as a reminder that the 'agricultural' boundaries between parishes (ie the boundaries of the holdings that comprise the parish summaries) may well differ from those of the civil parishes, the latter have been omitted. It is true that this is contrary to psychological theory, which indicates that the eye is better able to interpret the shapes of areas on maps when these are bounded by sharp lines; but on balance it seemed better to omit them, for it is always possible to replace them by superimposing a transparent overlay prepared from Fig 1 on any particular distribution.

Because of the variation in the sizes of parishes and because of the different levels of generalisa-

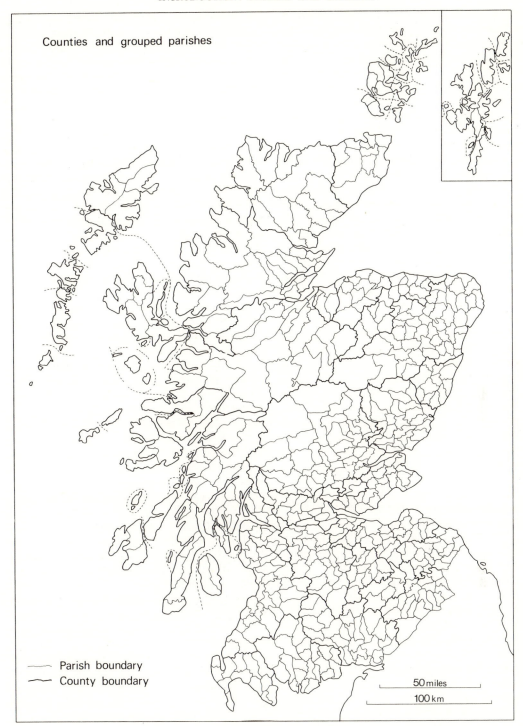

Counties and grouped parishes

Parish boundary
County boundary

50 miles
100 km

Fig 1

tion represented by parish summaries, maps of ratios and densities may well suggest that farming is more variable in some parts of the country, where parishes are small, than in others, where they are large, although such differences are only apparent and would disappear given the same degree of generalisation. It is thus desirable to standardise the parish data in order to eliminate the effects of differences in parish size on the degree of generalisation represented by the parish summaries. Such standardisation would be possible only by referring back to the original data to provide summaries of areas of approximately equal size, such as 10km squares of the National Grid (and only then if the locations of the farms were known). Since it was also desirable to amalgamate some of the smaller parishes, which could not be represented at the scale of the printed maps and since it was not possible to recombine the data to provide summaries of equal areas, a compromise solution was adopted in which summaries of the smaller parishes were amalgamated with those of their neighbours where this was possible without making an already large parish much larger. The target was a minimum size of parish of 15,000 acres (6,000ha), though it was not always possible to find a suitable partner for all the smaller parishes. In effecting such amalgamations, the principal criterion was size, but attention was also paid to similarity of land type, for there seemed to be no point in amalgamating parishes which were known to be unlike in this important respect. As a result of such amalgamations, the original 891 parishes were reduced to 557 combined parishes (Fig 1).

It is also worth noting that, as a result of the discordance between civil and agricultural parishes, it is not strictly correct to compute the relationships between the data contained in the parish summaries and those derived from other sources, such as the records of the Meteorological Office, or the Soil Survey or the population census, or even the total area of the parish, although in practice any errors are likely to be small once the parish data have been standardised.

The choice of the cartographic method to be used to represent the data was influenced partly by the characteristics of the data themselves, but

also by the shape of Scotland and the distribution of low ground within the country. A sound case can be made on both cartographic and statistical grounds for mapping agricultural distributions by means of choropleth maps; for their construction involves a degree of generalisation or smoothing, which lessens the effects of the known deficiencies of the data and yet retains a large part of the quantitative information which the agricultural census is capable of providing. It is true that the shading of each administrative area suggests a degree of correspondence between area and data which is unlikely to exist; but the omission of boundaries lessens this impression and the division of data into classes, each of which is shaded in respect of the agricultural characteristic being mapped, greatly facilitates comprehension of the data. By contrast, dot maps do not retain the quantitative information provided by the census (since, in practice, the dots cannot be counted) and give only a very generalised impression. In a similar agricultural atlas of England and Wales, which was also based on census data, nearly all the agricultural distributions were shown by choropleth maps, the chief exception being a set of maps of highly localised horticultural crops, for which circles proportional to area were used; when the distributions being mapped related to agricultural activities which took place only on improved land, as the great majority did, the major areas of rough grazing were left unshaded and were further identified by a thick boundary line. Such a solution was impracticable in Scotland, where 74 per cent of the agricultural land is classified as rough grazing, and where most of the improved land outside central Scotland lies near the coast. All distributions are therefore shown as dot maps, the use of choropleth maps being almost entirely confined to the display of data showing the relationship between different agricultural parameters, eg ratios of different classes of stock or crops per 100 acres of agricultural land.

In locating the dots on such distribution maps, few assumptions have been made about the likely location of particular crops or classes of livestock within each parish; for such assumptions may lead to circular arguments, in which dots are located on the basis of some assumed relationship and that relationship is then confirmed by reference to the map. With the exception of beef cattle and sheep, dots have been distributed evenly within the area of improved land, or more correctly, since this has not been accurately mapped, in those parts of parishes not occupied by rough land. Dots representing sheep have been uniformly distributed throughout the total area of agricultural land, whereas those representing beef cattle have been excluded from the higher areas nominally under rough grazing. It has not been possible to take account of the distribution of woodland, which is both changing and fragmented, or of urban land (except for the four counties of cities—Aberdeen, Dundee, Edinburgh and Glasgow). As far as possible, the dots representing the data in each parish summary have been placed within the parish boundary, although for the reasons already noted, some discretion has been exercised, particularly on small-scale maps showing the distribution of highly localised and specialised crop and livestock enterprises. To facilitate interpretation of the distributions, the land over 500ft (150m) is shown on many of the dot maps by means of a light stipple; it must not be inferred, however, that this elevation is of equal significance throughout the country, for the upper limit of cultivation is lower in the west than in the east.

In interpreting the choropleth maps, it must be remembered that data for many of the larger parishes in the uplands frequently relate to small numbers and only to part of the parish; all that is being asserted on such maps is that, on average within the general area represented by the parish in question, this particular relationship prevails.

Class intervals have been chosen by inspection of cumulative frequency graphs and by reference to a variety of statistical measures, including means, quartiles and standard deviations, but the choice of values was moderated by some consideration of what could be most readily interpreted by the general user. As far as possible not more than six classes of shading have been used, with class intervals at constant points along a scale of either fives or tens; to facilitate comparison, the same scale has been used for maps of similar crops or classes of livestock wherever this was possible. Some attention has also been

given to 'mappability', that is, the extent to which the areas under a particular category of shading fall into fairly simple patterns which can be readily appreciated by those unused to reading statistical maps.

To have undertaken the many complex calculations used in the preparation of these maps by hand would have delayed its preparation even further, and its production has been possible only with the help of a computer. Compilation of the maps was also facilitated by printing the results of the computations not only in tabular form but also in their true spatial relationship by reference to a grid of coordinates, each parish result being printed at the approximate centroid of the parish. This approach has subsequently been developed into a complete computer mapping system, but this was not available at the beginning of this project and, by the time such a mapping program was ready, a large proportion of the maps had already been drawn by hand. In the preparation of such an atlas in the future it will be possible to produce maps suitable for block-making direct from the line printer or from the graph plotter, thus eliminating much tedious cartography and reducing the risk of error; for the more frequently the data are handled, the greater the risk that errors will occur.

TABLES

To supplement the maps, tables are included to provide some broad quantitative indication of the importance of the data being mapped. These tables relate either to Scotland as a whole or to the five regions into which Scotland is divided by the Department of Agriculture and Fisheries for Scotland (Fig 1). These are groups of counties, viz: Highland (Argyll, Inverness, Ross and Cromarty, Sutherland, Zetland); North East (Aberdeen, Banff, Caithness, Kincardine, Moray, Nairn, Orkney); East Central (Angus, Clackmannan, Fife, Kinross, Perth); South East (Berwick, East Lothian, Midlothian, Peebles, Roxburgh, Selkirk, West Lothian); and South West (Ayr, Bute, Dumfries, Dunbarton, Kirkcudbright, Lanark, Renfrew, Stirling, Wigtown). Reference to these regions is indicated by the use of initial capitals, eg North East Region, the

North East; where less precise regional descriptions are intended, lower case is used, eg north-east Scotland (Buchan and adjacent areas).

THE ARRANGEMENT OF TEXT AND MAPS

Although this book is primarily an atlas and pride of place is consequently given to the maps, an attempt has been made to provide a brief commentary on the maps, though coverage is inevitably somewhat uneven. The maps and text are arranged in seven sections. The first describes those features of the natural environment which are important for agriculture and for which mappable data exist, a constraint which leads to an underrepresentation of soils and of those aspects of climate which particularly affect farming in Scotland, such as the onset of the growing season and the variability of weather. This section concludes with a discussion of rent, which serves to link it to the following section on the man-made contribution to farming, with a particular emphasis on maps of employed labour, machinery, markets and agricultural holdings; this coverage, too, is limited by available data, and important inputs, such as fertilisers and feeding stuffs, are omitted, while information on machinery is available only by counties. The third section is devoted to land use, and as in other atlases of agriculture in Great Britain,

exaggerates the importance of cropping because this information is available in greater detail than that for livestock; even so, no mappable data for yields or varieties exist and the maps show only acreages. The following section on livestock is appropriately the longest in the book, with pride of place given to beef cattle and sheep, the latter, not so much because of their importance in the agricultural economy of Scotland as for their distinctive geographical characteristics; it should be noted that nearly all these maps show summer distributions which may differ significantly from those in winter. The fifth and sixth sections represent attempts to integrate the information in the preceding sections by examining respectively agricultural enterprises and types of farm; these sections are inevitably more speculative since they require decisions about what constitutes an enterprise and a farm type, concepts which, despite general impressions to the contrary, are not self-evident. The final section provides an historical perspective by displaying the distributions of the principal crops and livestock in 1870 and in 1938, and by examining trends over the century since 1870, thereby revealing both the stability of the main components of Scottish agricultural geography and the detailed changes which have altered the balance between individual crops and classes of livestock.

1

The Physical Basis of Agriculture

THE physical conditions under which farming takes place are a major determinant of the choice of farming systems, not so much by prohibiting the selection of particular enterprises as by affecting the yields of crops and livestock (or livestock products) or the costs of producing them, and hence their profitability vis-à-vis other enterprises. The availability of statistical data on these physical conditions is very uneven: a great deal is known about relief, which has been recorded for more than a century on the maps of Ordnance Survey, rather less about climate (other than rainfall) and least about soils, particularly for Scotland as a whole. Knowledge of the precise effects of differences in relief, climate and soils on the performance of different crops and classes of livestock is even more patchy, and all that can be attempted in this section is to indicate the main characteristics of each of these aspects of the physical environment and how it varies throughout Scotland, and then to ask the reader to bear these facts in mind in interpreting the maps showing individual crops and classes of livestock. Since

relief is an important factor in both local climate and soil type, it will be discussed first.

RELIEF

Scotland contains the highest elevations in the British Isles, reaching 4,406ft (1,343m) in Ben Nevis, and has a higher average elevation than any other of the constituent countries. Estimates made for the Royal Commission on Coast Erosion and Afforestation in 1909 show that some 2·6 million acres, or 14 per cent of the country, lie above 1,500ft (450m), the county with the highest proportion being Perthshire, with 32 per cent; and, according to G. P. Glentworth, some two-thirds of Scotland lie above 1,000ft (300m). Fig 2 shows the generalised distribution of relief in Scotland and confirms the impression of the small proportion of lowland. Apart from the proportion under trees, which is probably less than a tenth, most of the land above 500ft (150m) is rough grazing or other rough land, though the upper limit of cultivation reaches much higher on the more sheltered, less maritime valleys of eastern

Scotland than it does on the west coast; in the western Highlands, where rough land may reach almost to sea level, there is little cultivated land above 100ft (30m), whereas in eastern Scotland the upper limit of improved land reaches as high as 1,000ft (300m) in places. This varying significance of particular elevations in different parts of the country, a reflection of differences in exposure and in maritime influences, and of the different sizes of the upland masses, must be borne in mind in interpreting the maps of agricultural distribution in this atlas, for the larger the upland, the higher the upper limit of cultivation is likely to be. Little of the land above 2,000ft (600m) is of any agricultural significance and there is, in any case, a tendency for agriculture to retreat from the higher tracts, leaving them to the red deer which already have exclusive use of the highest grazings.

Apart from the central lowlands, which are themselves interrupted by a line of hills derived from igneous rocks, stretching from the Renfrew Hills in the west to the Sidlaw Hills in the east, and by the watershed between the Forth and the Clyde drainage systems, low-lying ground is largely confined to the coasts of Galloway and the Solway lowlands, the Merse (or lower Tweed valley), Buchan and the plain of Caithness. In the last, as in many of the islands, other factors, such as surface wetness and deep peat, tend to offset the advantages of low elevation. On the west coast of the Highlands only small patches of the relatively small extent of lowland are suitable for cultivation, generally on raised beaches or river terraces; most of the remainder is either too rugged or covered with peat, and so is suitable only for rough grazings. Even within the central lowlands, the extent of level or near-level land is less than the map suggests, for much is covered by drift deposits and is hummocky and uneven. Within the uplands there are obvious contrasts between the Southern Uplands and the Highlands; only some 6 per cent of the former is above 1,500ft (450m) and the landforms are generally smooth, with extensive tracts of rolling country. The Highlands are not only higher, but are also more rugged, with extensive outcrops of bare rock or boulder-strewn surfaces. There is also an obvious contrast between the eastern and western Highlands.

In the former, especially in the Cairngorms, there are extensive tracts of fairly level land at high elevations; the western Highlands, on the other hand, are much more dissected, so that there are few upland plateaux, and mountains often rise abruptly from coast or valley.

Relief as such is not generally a major obstacle to agriculture in Scotland. Its main influence is felt indirectly through the modification of climate, notably in relation to both amounts of rainfall and temperature regimes; for the steep lapse rate leads to a rapid shortening of the growing season with increasing altitude. The disposition and character of the present landforms thus result in much sharper regional and local contrasts in climate than would otherwise be the case. Little land that is otherwise capable of cultivation is too rugged or too steep to be used for agriculture, and paradoxically, it is often the flat or gently sloping land that is marginal, because it is too poorly drained in its natural state to be farmed. In the lowlands there are still quite extensive tracts of peat that have not yet been drained, and the flat floors of the valleys in the uplands are often ill-drained haughlands (though the coarseness of the material filling the valley is also a factor here). In the uplands, A. G. Ogilvie has drawn attention to what he called 'debatable lands', lying between 400 and 1,200ft (120–360m) and with a slight gradient; these, too, are handicapped by poor drainage. Large tracts of upland rough grazings below 1,000ft (300m) were cultivated in the past, as the evidence of cultivation ridges shows; their adandonment is largely a reflection of climatic change and of a move from subsistence farming, but they show that there are no insuperable topographic obstacles to the cultivation of such land.

SOILS

No detailed soil map of Scotland is yet available, though surveys by the Soil Survey of Scotland are in progress in different parts of the lowlands and special surveys have been done elsewhere, as on Mull. Figs 3, 4 and 5 display, in broad outline, the main feature of Scottish soils and have been prepared by the Soil Survey; because the maps are so generalised, no significance

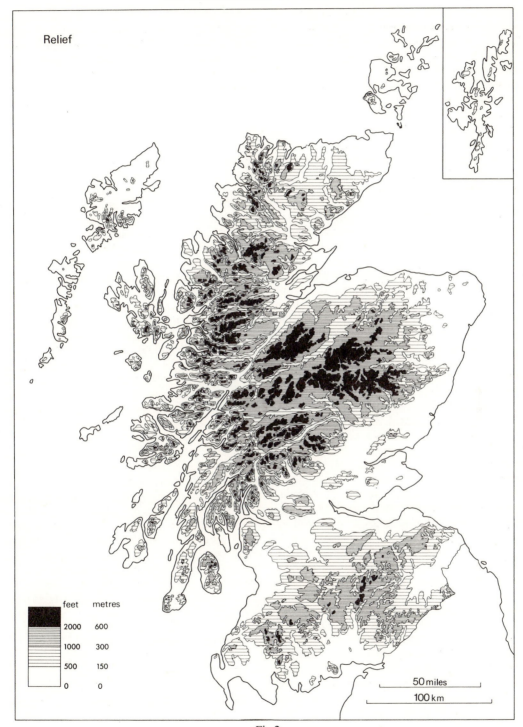

Relief

feet	metres
2000	600
1000	300
500	150
0	0

50 miles

100 km

Fig 2

should be attached to the precise location of any of the boundaries shown.

The distinctive features of Scottish soils are their youth and their generally low level of natural fertility. They have been formed since the retreat of the Pleistocene ice sheets some 10,000 years ago and are mainly derived from drift deposits, which in turn are largely the product of acid rocks. Those in and around the Highlands tend to be coarse textured, whereas those in the lowlands are generally of finer texture, especially in the western half of the central lowlands. For this reason, and because of the distribution of rainfall and the balance between precipitation and evaporation (Figs 7 and 11), many Scottish soils are imperfectly drained. The highest land, above 2,500ft (750m), is either bare rock or covered with skeletal soil; about a tenth of the upland is under blanket peat (ie more than 12in or 305mm thick) and most of the remainder is covered with peaty podsols in which thin ironpans underlie peaty horizons. On the lower ground the principal contrast is between the eastern and western half of the central lowlands; the finer texture and heavier rainfall of the west lead to more poorly drained soils than further east (Fig 3).

The main textural differences, shown in Fig 4, have already been indicated. Loams characterise the Lothians and the Merse, as well as the lowlands of Angus, Kincardine and Perth, ie the principal arable areas. In Buchan and on the north-west and south-east coasts, stony loams and sands abound, whereas the soils of much of the central lowlands are clays and clay loams. Peats, stony and peaty soils and bare rock occupy most of the uplands.

Fig 5 shows the generalised distribution of the main genetic soil groups. The areas shown as loams on the map of texture (Fig 4) are classified as mainly brown forest soils, though they are generally comparatively poor in bases. Those derived from coarse-textured materials in Buchan and elsewhere around the coasts are podsols and brown acid soils. Gleying is characteristic of the poorly drained fine-textured soils of central and west-central Scotland, and the types shown here include both gleyed brown forest soils and surface water gleys. Skeletal soils, peaty podsols and peats are the principal soils of the uplands.

Against this background, it is not surprising that first class land should be rare in Scotland (Fig 6). Fig 6 and Table 1 are derived from the work of the Land Utilisation Survey, which still provides the only complete assessment of agricultural land quality for Scotland. It is true that staff of the Department of Agriculture and Fisheries for Scotland prepared detailed maps of agricultural land quality of most of lowland Scotland in the 1940s, in which each field was classified on a six fold point scale, but these have never been synthesised into a single map; more recently, the Macaulay Institute has embarked on the preparation of land capability maps (conceived from the viewpoint of agriculture), but only a few of these have so far appeared. The Land Utilisation Survey's classification is a subjective one based on available map evidence on land utilisation and physique, and Fig 6 is highly generalised. First class land is that capable of intensive cultivation and of being worked at all seasons. It is included within the category of good land, defined as highly productive under good management and comprising level or gently undulating land with favourable aspect, not too elevated, and with deep, well-drained soils. Medium quality land is that which is only moderately productive, even under good management, because of the operation of one or more adverse factors, eg excessive height or steepness, unfavourable aspect and shallow or poorly drained soils. Poor quality land is that affected by the extreme operation of one or more of these factors. First class land is largely confined to tracts of brown forest soils in the Merse, the Lothians and the lowlands between Perth and Stonehaven. Most of the remaining lowlands are classified as predominantly good land, medium-quality land being largely confined to the upland fringe. It must, however, be appreciated that the prevalence of drift deposits throughout lowland Scotland leads to a highly variegated pattern of land quality which is lost through generalisation at this scale.

TABLE 1

Percentage Proportion of Land in Different Classes

First class	Other good	Medium	Poor
2·1	18·7	15·1	63·5

Source: Land Utilisation Survey

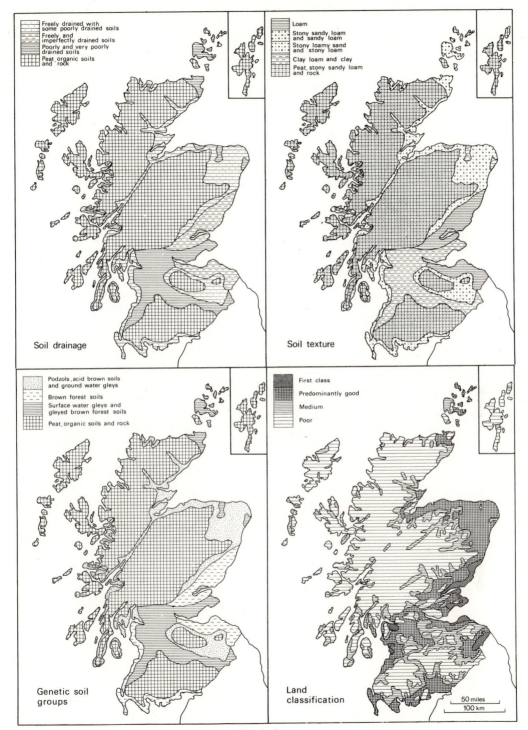

Figs 3–6

CLIMATE

Climate is the most important physical factor affecting agricultural activity in Scotland, although it is itself much modified by the range of relief and the disposition of land forms, by proximity to relatively warm seas and by the deep interpenetration of land and sea. Differences also arise from the fact that Scotland extends over some six degrees of latitude or more than 400 miles (640km), a greater latitudinal extent than England. There are thus contrasts between north and south, east and west (reflecting the balance between maritime and continental influences), coast and interior, and lowland and upland. Because of the steep lapse rate, small differences in aspect and exposure can produce marked climatic contrasts. Year-to-year variability, too, is a characteristic feature of many aspects of Scottish climate, and one that is of considerable importance to the farmer.

Rainfall (here intended to mean precipitation in any form) is the best documented and recorded of all climatic parameters, for there are some 1,250 rainfall recording stations; their distribution, however, is very uneven, especially on higher ground, and interpretations of rainfall distribution above 1,000ft (300m) must be made with caution. The mean average rainfall is just over 53in (1,346mm) for Scotland as a whole, but varies from under 30in (750mm) along much of the east coast and the Merse to over 100in (2,500mm) and possibly as much as 200in in the western Highlands. There are two main factors affecting the distribution of rainfall, elevation and location, for rainfall increases with height and, for any given height, from east to west; precipitation is both orographic and frontal, derived mainly from the sequence of eastward-moving cyclones which characterise much of Scotland's weather. Fig 7 and Table 2 give some indication of the relative importance of different amounts of annual rainfall.

TABLE 2

Percentage of Scotland with Different Annual Rainfalls

Rainfall in inches	under 25	25–30	30–40	40–50	50–60	over 60
Percentage of area	1	6	29	25	17	28

Source: E. C. Bilham, Climate of the British Isles

Along most of the western coastlands and the islands rainfall is between 40 and 60in (1,000–1,500mm) and increases sharply with elevation to over 60in, exceeding 100in (2,500mm) on large areas of higher ground in the western Highlands. The eastern Highlands, by contrast, have less than 60in and values fall to between 30 and 40in in the major valleys, such as those of the Spey and the Don. In southern Scotland the lower elevation of the Southern Uplands and north-east/south-east grain of the major features of the relief lead to less marked contrasts between east and west; for the north/south orientation of much of the Highland relief acts as a barrier. Similar differences between east and west exist in the lowlands, both along the coasts and in the central lowlands, where the Clyde–Forth watershed corresponds approximately with the 40in isohyet.

These differences in precipitation are also accompanied by differences in the number of rain-days, ie days on which at least 0·1in (2·5mm) of precipitation falls. These range from under 175 days on parts of the east coast to more than 250 days in the western Highlands, although the relationship between rainfall and number of rain-days is not simple; thus the Hebrides, Orkneys and Shetlands have a much larger number of rain-days than might be suggested by a consideration of their annual rainfall alone, and there can be quite marked local differences in total rainfall between the summit and foot of a mountain and yet little difference in the number of rain-days.

A proportion of the annual precipitation comes as snow, which may fall on twenty to thirty days in northern lowlands, but on less than twenty in the south—figures that rise by about one day for every 50ft (15m) to about 1,000ft (300m) and more rapidly thereafter. Much of the snow falling on low ground does not stay long and snow lies on average for only five days a year in the south-west, for only ten on the east coast and for about twenty days in the interior lowlands. At high elevations in the Highlands, by contrast, snow lies on average for more than 100 days, and short-term records for Ben Nevis gave an average of 170 days.

These figures generally represent long-term averages, taken in the main over a period of

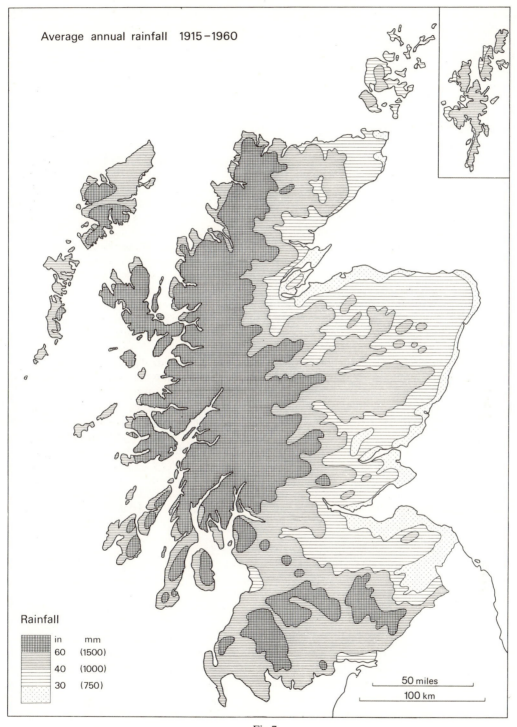

Average annual rainfall 1915–1960

Rainfall

in	mm
60	(1500)
40	(1000)
30	(750)

50 miles

100 km

Fig 7

forty-five years between 1915 and 1960, but such values may vary by more than 50 per cent from year to year, variability being generally greatest where rainfall is lowest, notably in south-east Scotland. Such variability, as measured by the coefficient of variation, ranges from between 8 and 12 per cent in the west to between 16 and 20 per cent in the east.

For most agricultural purposes, it is the amount and distribution of summer rainfall which is important to farmers, though the depth and duration of snow are aspects of concern to those farming hill country. As Fig 8 shows, there is a general resemblance between the maps of annual and summer rainfall, but the differences between east and west are less marked in the latter. This is because of regional differences in the distribution of rainfall (Fig 9); in the west, most rain falls in the winter half of the year (October–March) and the proportion tends to be greatest where rainfall is heaviest. Thus, at Oban, 24·2in (615mm) fall in the summer six months and 32·3in (820mm), or 58 per cent of the annual total, fall in the winter half. In the east the rainfall is more evenly distributed, with a tendency to a maximum in the summer half of the year; at Turnhouse Airport, Edinburgh, 14·2in (361mm) fall in the summer six months and 12·7in (323mm), or 47 per cent, in the winter months.

In a British context, Scotland has a cool climate, though it is warm for its latitude in winter, a fact partly obscured by the relatively high elevation of much of the country and the steep lapse rate. In winter, the isotherms of temperature reduced to sea level run north/south and mean temperature increases from east to west; but in summer, when warmth is important for crop growth, they run east/west and the extreme south of the mainland is some 5°F (3°C) warmer than the north. The coast is relatively warm in winter and cool in summer, a fact of some importance where so much low land is coastal and spring tends to be late, and where temperature is a restraining factor on crop production. Since growth tends to increase with temperature, a useful indicator of temperature is accumulated temperature (Fig 10), which is measured in day-degrees and integrates the

number of days on which the mean temperature exceeds some base temperature (here 42°F, or 6°C which represents the approximate threshold at which plant growth begins in many indigenous species), and the number of degrees above this minimum. Virtually all the uplands record fewer than 1,000 day-degrees Centigrade, but values in the lowlands range from 1,121 in Wick to 1,344 in Inverness and 1,504 in Dumfries.

Temperature and rainfall do not, however, act in isolation, for precipitation is offset to some degree by evaporation. Potential evaporation is a computed measure of the loss of water by evaporation from a green crop covering the ground and with no limitation on the supply of water to the roots. More than three-quarters of this water loss occurs in the summer six months (Fig 11). In winter, any deficiency in soil moisture is quickly made good and subsequent surpluses are lost through run-off or percolation; at some time in summer most soils tend to dry out. The extent of any deficit or surplus can be established by comparing Figs 8 and 11. In the east such deficiencies may justify the application of irrigation water to valuable crops, especially in late spring and early summer.

There are, of course, many other aspects of climate which affect agriculture. Late frosts occur at progressively later dates away from the coast and with elevation, the last spring frost occurring on average before 15 April on the west coast and before 1 May over most of the east coast. Wind is a particular feature of the Scottish climate, especially in western and northern areas and with increasing altitude where, in conjunction with low temperatures and driving rain or snow, it may subject animals to severe stress and pose difficult problems for hill farmers. Average wind speed in Shetland, for example, is twice that in sheltered valleys in the Highlands, and the widespread distribution of shelter belts is recognition of the importance of wind as a hazard. Many of these factors, viz accumulated temperature, potential evaporation, frost and exposure, have been integrated into a series of climatic maps by staff of the Macauley Institute, and Table 3 gives some representative figures for various parts of Scotland.

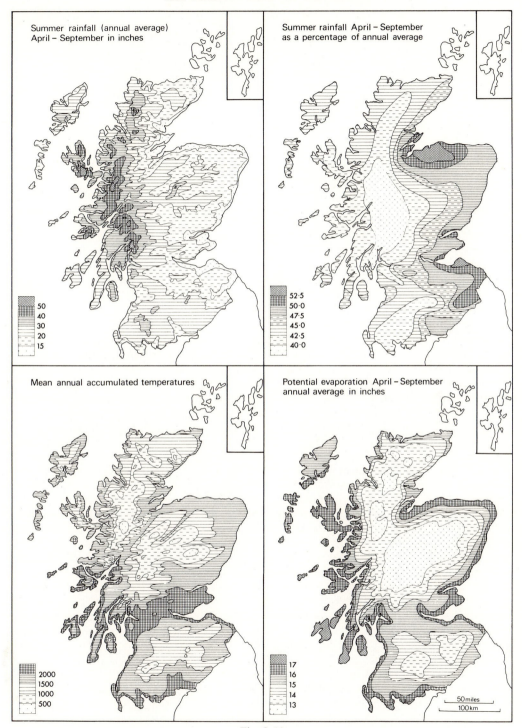

Figs 8–11

TABLE 3

Selected Meteorological Data

| | Mean temperature: | | Mean growing season days | Daily mean sunshine hours | Mean annual rainfall ins | Average number of days with snow lying days |
	Warmest month °F	Coldest month °F				
Wick	55	39	237	3·7	30	15
Dalwhinnie	55	33	203	2·9	47	58
Braemar	56	34	186	3·1	37	62
Dundee	59	37	234	3·7	27	15
Tiree	57	41	310	4·0	45	3
Renfrew	59	38	243	3·3	41	11
Eskdalemuir	56	35	199	3·3	62	25

Source: Meteorological Office

SEASONALITY

The agricultural significance of meteorological phenomena depends on their interpretation by farmers. While climate sets ultimate limits to all forms of agricultural production it is very rare that these limits are closely approached, let alone reached; for economic constraints, as reflected in higher costs of production or lower returns, lead to changes in the nature of agricultural production long before this point is attained or, where no alternative is possible, to the abandonment of agricultural production altogether. In any case, owing to the great variability of climate from year to year, farmers have to make judgments of the likelihood of particular meteorological events occurring.

Climate not only affects the nature of agricultural production; it is also responsible for one of the most distinctive features of agriculture, its seasonality. It is true that the cyclic nature of agricultural production is in large measure biological, but biological cycle and climatic cycle interact, and the more seasonal the climate the more closely constrained is the sequence of events in the agricultural year. Scotland, lying towards the northern limits of the temperate zone and having sharp climatic gradients both with increasing elevation and from the coast inland superimposed on a latitudinal zonation, exhibits quite considerable regional contrasts in the calendar of agricultural events. Unfortunately, there are few firm data to substantiate this assertion and, to remedy this deficiency, help was sought from the agricultural

inspectorate of the Department of Agriculture and Fisheries for Scotland. These inspectors and their colleagues are distributed throughout Scotland and have considerable first-hand knowledge of the agriculture of the areas for which they are responsible. They were asked, among other things, to record on maps of their areas the average dates on which selected events in the farmer's year took place; these maps were then collated to provide a national picture of the situation throughout Scotland. It must be recognised that the data have a number of limitations: they are impressions, rather than objective measurements according to defined criteria, and there were inevitably discordances at the boundaries of adjacent maps compiled by different inspectors. They also indicate average conditions over a number of years and, given the variability of the Scottish climate, it is likely that the situation in a single year may differ considerably from that revealed by this survey; it is even possible for the latitudinal sequence to be reversed in some years. Nevertheless, it is on such assessments of the likelihood of particular events occurring that the farmer has to plan the long-term investment on his farm and decide the kinds of enterprises he wishes to practise.

A selection of the maps prepared from this survey is shown in Figs 12–15. They resemble, in their nature, the maps of floral isophenes prepared by botanists, which record on the basis of observation of selected species, the dates at which they flower throughout the country;

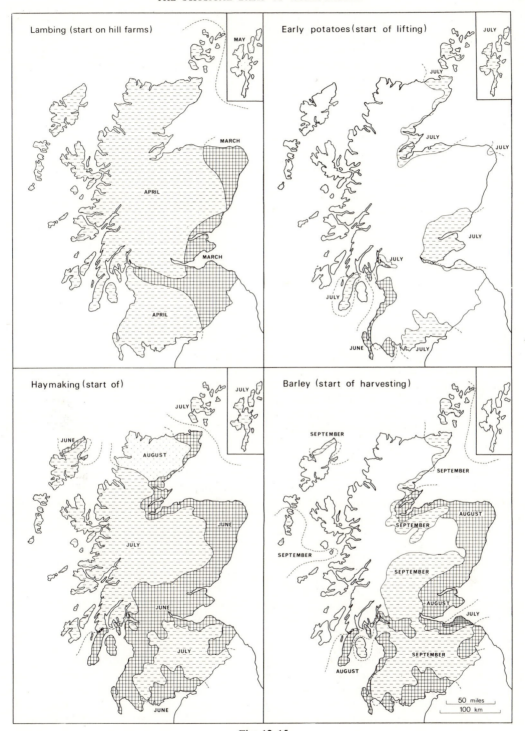

Figs 12–15

these maps are thought to be the first of their kind prepared for Scotland as a whole (though some agricultural calendars have been made, and in one study the sequence of events in each field in every month has been prepared for selected farms). They are a reminder of the fact that each of the agricultural distributions in this atlas represents only a snapshot, an instant view of dynamic processes, in which ground is prepared, crops sown and tended and finally harvested, or livestock born, bred and fattened, or transported elsewhere for fattening. The events shown in these maps should be read in conjunction with the appropriate agricultural maps.

The onset of lambing is related to both breed and environment. Most breeds of sheep kept in Great Britain lamb in spring and there is generally a sequence of progressively later starts with elevation and exposure, though much will depend on farm practice: Fig 12 shows in very generalised form the pattern of lambing on hill farms. Early potatoes are mainly grown in the lowlands of Angus, Fife and Kincardine, but the earliest harvests occur along the sandy coastal fringe of Ayrshire and south-west Scotland (Fig 13). According to Fig 14, the hay harvest begins in June in most of lowland Scotland, but does not generally start until July in most uplands and along the west coasts and in the islands. The barley harvest, as shown in Fig 15, generally begins in August in the lowlands, except in the Lothians where the start is in July; in most western areas, by contrast, it does not begin until October. These maps are included, not only for their intrinsic interest, but also as an indicator of what a comprehensive agricultural atlas might contain.

RENT

Rent provides a convenient link between the natural resources which have been considered in this section and the man-made framework within which decisions are made, ie the size and shape of farms, their layout and the capital which has been invested in buildings and other fixed equipment. It also reflects to some extent the location of farms in relation to the main centres of population where the demand for agricultural land, whether for transfer to other uses or for farming by those who do not depend wholly on farming for their living, tends to increase land values.

Data on the rent of agricultural land are not now collected from all holdings. However the Department of Agriculture and Fisheries for Scotland does monitor rent levels on a sample of about 1,000 tenanted or partly tenanted farms, which form part of a random sample of more than 2,000 full-time holdings, and publishes information on percentage changes in rents (though not normally absolute values) in the annual volumes of *Scottish Agricultural Economics*. The last year in which data were collected from all holdings was 1960, and it is these data which have been used in the compilation of Fig 16. Since more than half the farmland in Scotland is farmed by those who own it, data on rents are available for only a minority of farmland, and estimates of rents on the remainder are provided by the gross annual value of land, which is also recorded. These two sets of figures are broadly consistent, the average true rent per acre in some counties being higher and that in others being lower than the average gross annual value per acre. The values mapped in Fig 16 were obtained by adding the total rent in each parish to the total gross annual value for owner-occupied land and dividing this sum by the total acreage of crops, grass and rough grazing. Because of rising rents in the period since 1960 (especially since 1970), this map must be viewed as an indicator of relative importance rather than one of absolute values.

The average rent of agricultural land in Scotland in 1960 was £0·39 per acre, with county totals ranging from £1·52 in Fife to £0·03 in Sutherland, and parish averages ranging from £4·11 to less than £0·01. Fig 16 shows that average rents in 1960 (before the 1958 Agriculture Act had had time to affect rents greatly) exceeded £2·00 per acre in the Lothians and along the coastal lowlands from Fife to Kincardine, as well as around Aberdeen, Ayr and Glasgow; average rents exceeding £1·50 per acre were also to be found mainly in these areas. In the western half of the central lowlands and in the hill country of north-east Scotland rents were generally between £0·50 and £1·00 per acre; on the fringes of the Highlands, in the Southern Uplands, Caithness, Orkney and the

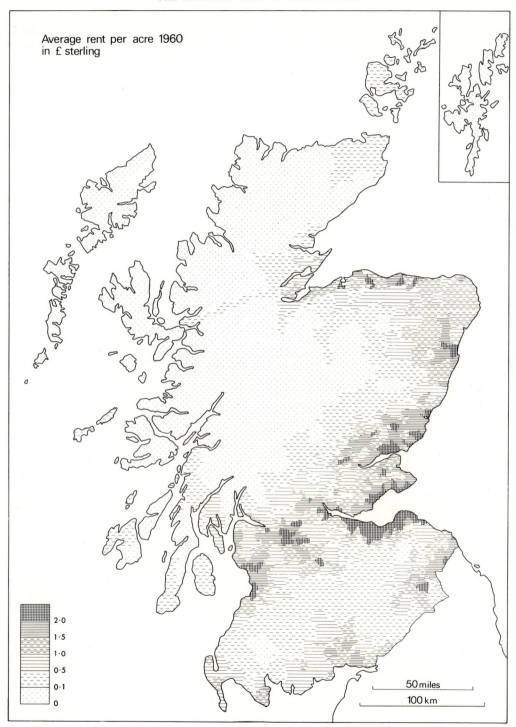

Average rent per acre 1960
in £ sterling

2·0
1·5
1·0
0·5
0·1
0

50 miles
100 km

Fig 16

islands of the south-west, between £0·10 and £0·50; and in the Highlands and the remaining islands, below £0·10. These averages, of course, conceal a wide range of values; in 1962–4, for example, rents on cropping farms ranged from under £0·50 per acre to over £10·00.

There is little published information relating rents to farm size, type of farming or land quality, each of which can be expected to have some influence upon the level of rents. According to an investigation by the Department of Agriculture of rent levels on sample holdings in relation to acreages of crops and grass alone (a procedure which clearly exaggerates rent levels on farms with much rough grazing), rents were highest on cropping farms in 1962–4, averaging £3·05 per acre, compared with £2·35 for dairy farms. On this basis of comparison there is little evidence of the decline that might be expected with increasing size of farm; indeed, in north-east Scotland, where small farms often have poor buildings, and on cropping farms, where there was competition for farms to rent, there is even an opposite trend.

Rent represents only about 6 per cent of farm expenditure in 1965, but it is nevertheless an important outlay for farmers because it is a fixed charge which has to be met.

2

The Man-made Framework of Farming

WITH the exception of rent, the maps in the preceding section have been concerned with the physical conditions under which farming takes place and these exercise a major influence on the character of Scottish agriculture. Nevertheless, farming is also affected by the size of farms, the units of decision-making, and by the resources that are used in conjunction with land and without which no agricultural production could take place: the amount of land which the farmer controls, the labour and machinery which he employs, the feeding stuffs and fertilisers which he purchases to increase the output of his farm, and his capital investment in fixed equipment whereby he escapes some of the limitations of his natural environment. Unfortunately, information on many of these factors is available only for Scotland as a whole, sometimes for the five regions and occasionally for counties, and little is suitable for mapping. There is thus a danger of giving a misleading impression by displaying only those aspects for which there are mappable data. Such an approach will tend to emphasise land at the expense of other factors even though it is one of the least important components of farmers' expenditure. In 1965, for example, labour accounted for some 31 per cent of all farm expenditure in Scotland (excluding feeding stuffs and purchased store stock) and machinery and fuel 29 per cent, compared with only 9 per cent for rent (or 21, 19 and 6 per cent respectively of all expenditure). Although the proportions vary somewhat throughout the country and with different types and sizes of farm, labour is always the most important item of expenditure, followed by machinery, with rent accounting for a much smaller proportion. This account will be largely confined to labour, machinery and farm size, though some attention will also be paid to tenure of farm, the outlets for agricultural produce and the movements which link these.

LABOUR

The most abundant data are those relating to farm labour, on which questions have been included in the annual agricultural census for over fifty years. Unfortunately, this information is only partial and to that extent misleading, since it relates to employed labour only and does not include the contribution of the farmer and his wife, although this is likely to be very important on the smaller holdings. It is true that the census now makes it possible to distinguish

23

between full- and part-time occupiers, but even among full-time occupiers there is a continuum between those whose energies are largely devoted to manual work and those whose role is almost entirely managerial.

The labour force so defined is divided in the census according to age, sex and the amount of time devoted to agricultural employment. For the construction of Fig 17, these various categories have been converted into full-time equivalents; for this purpose, a rough and ready measure of one unit for each full-time worker, one half unit for each regular part-time worker and one third unit for each casual and seasonal worker, as recorded on 4 June, has been used. More sophisticated factors could have been employed, varying with the sex and age of the worker, but since 51,522 of the total labour force of 63,761 in 1965 were full-time workers, and 47,142 of these were males, the effect of using a complex system of weights is likely to be small. It would be possible to estimate the total labour force, including the contribution of the farmer and wife (or husband) by computing the total labour requirements of the 'parish farm', as has been done in Fig 191; but this approach relies on assumptions that labour requirements are the same everywhere and that labour is used with equal efficiency on all holdings. The results of such computations are more properly used as a comparative measure of the intensiveness of farming rather than as absolute indications of the size of the labour force.

On this basis of conversion there were 56,549 labour units in 1965 (compared with 41,034 in 1972) an average of three for every thousand acres of agricultural land (a term used throughout this atlas to describe the total area recorded in the agricultural returns, and therefore including not only crops and grass and rough grazings, but also a small acreage of farm woodlands, farm roads and buildings). The statistical distribution is highly skewed, with a mean of 7·0 units, a standard deviation of 7·5, a maximum value of 10·13 and a minimum 0·00. Fig 17 shows how values are distributed throughout Scotland. As in many of the maps in this atlas,

Fig 17 shows an expected contrast between the lowlands and uplands and between the Highlands and the Southern Uplands. Throughout the lowlands there were five or more labour units per 1,000 acres, though values were highest in East Lothian, East Fife and the lowlands of Perth and Angus; values are also high in the Merse, the coastal lowlands of Ayrshire and around Glasgow. In the Southern Uplands values are typically between one and three units, but throughout the Highlands proper they were less than one. Values were generally low in the islands. This distribution is a reflection of both farm size and type of farming though to some extent these factors work in opposition; values are high where cash crops make a major contribution to farm income and farms are large, and also to a lesser extent in the dairying areas, where farms are smaller. Similarly, values are low both in the hill farming areas, where farming is extensive and farms are large, and in the crofting districts, where most holdings are part-time and little labour is employed. Apart from hill sheep farms, however, data by type of farm for Scotland as a whole show little variation in total expenditure attributable to labour, though labour costs per acre increase with the intensity of the farming system (Table 4). In nearly all instances the proportion of labour costs and the actual labour input increase with size of farm (as measured by standard labour requirements).

TABLE 4

Labour Costs in 1965

Type of farm

Hill sheep	Upland	Rearing with arable	Arable rearing and feeding	Cropping	Dairy
Percentage of inputs (excluding seeds and feeding stuffs)					
46	35	35	31	37	35
Labour costs in £s per acre					
0·45	0·92	4·74	7·59	11·38	8·26

Source: *Scottish Agricultural Economics*, Vol 17

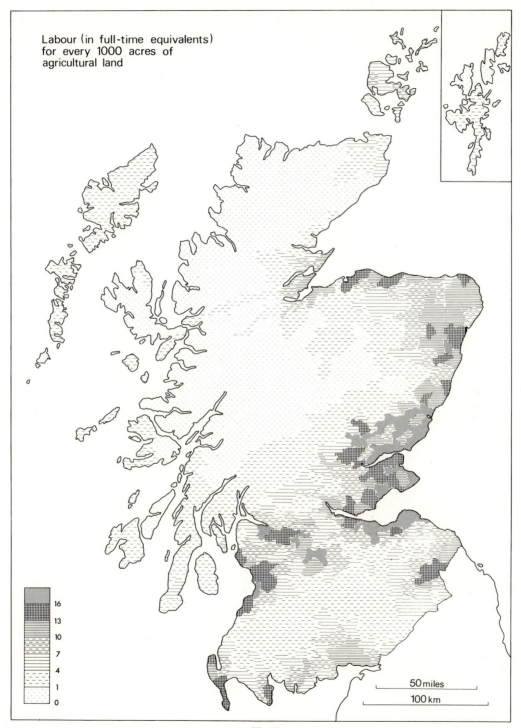

Labour (in full-time equivalents)
for every 1000 acres of
agricultural land

16
13
10
7
4
1
0

50 miles
100 km

Fig 17

Figs 18–21 show how the composition of the labour force varies throughout the country, although in absolute terms, measured per hundred acres, the distribution broadly resembles that in Fig 17; this is especially true of full-time workers, who accounted for 81 per cent of all workers (or 91 per cent of all labour units). Values for the proportion of full-time workers cluster closely around the mean, with a standard deviation of 10·8 and upper and lower quartiles of 76 and 88. Fig 18 reveals no clear pattern, with highest values in the Southern Uplands, in north-east Scotland and along the margins of the Highlands, and should be read in conjunction with Figs 19 and 20 and Table 5. The distribution of part-time and seasonal workers was much more skewed, with means of 9·4 and 9·9 per cent respectively and standard deviations of 6·3 and 8·0. Part-time labour was relatively most important in the western half of the central lowlands, in the main dairying areas, though there are scattered high values in the Highlands, especially on the west coast and on the islands (Fig 19); absolute densities are fairly uniform throughout the lowlands and very low elsewhere. Casual and seasonal workers, as recorded in June, also show an uneven distribution, being highest in the eastern part of the central lowlands and lowest in the uplands and in the north-east (Fig 20); though in absolute terms, such labour was most abundant in the lowlands of east and central Scotland. A sample survey in 1962/3 showed that 33 per cent of the casual labour used was employed between June and August, mainly on fodder crops and hay, and 44 per cent between September and November, when most was used on cash roots and grain. The use of seasonal labour on root crops is distinctive in that much of the labour employed is gang labour or people not otherwise gainfully employed; large numbers of women and children are particularly characteristic of the potato harvest. In interpreting these maps it should be noted that the distributions are even more patchy than they appear, for the percentage of holdings on which there is either casual or part-time labour is very low.

Both total labour and its distribution between full-time, part-time and seasonal labour are related to type of farming, though data are not available for mapping variations throughout the country. Table 5 shows how the labour force is distributed among different types of farm in Scotland as a whole and the composition, in relative terms, of this labour force; it refers to full-time farms only, which account for 93·1 per cent of all labour, and should be read in conjunction with the type of farming maps in Section 6. Hill sheep farms and intensive farms represent the extremes, but both cropping and dairy farms, which together account for over half the labour force, have below average proportions of full-time workers.

The last aspect of the labour force to be map-

TABLE 5

Farm Labour and Type of Farm in 1965

Type of farm

	Hill sheep	Upland	Rearing with arable	Arable rearing and feeding	Cropping	Dairy	Intensive	All
Percentage of all workers on each type of farm								
	4	10	14	7	22	33	8	100
Percentage of all workers								
Full-time	91	83	86	88	83	82	72	83
Part-time	5	8	6	6	7	10	13	8
Seasonal	5	9	8	7	11	8	15	9
All	100	100	100	100	100	100	100	100

Source: *Scottish Agricultural Economics*, Vol 17

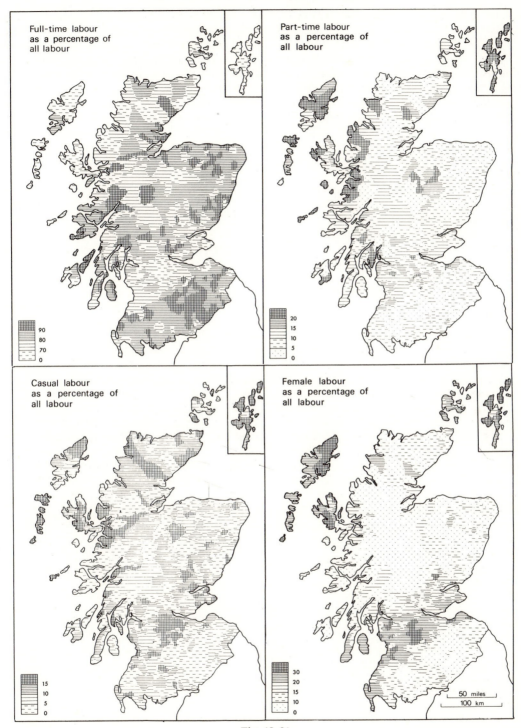

Figs 18–21

ped is the proportion of female labour (Fig 21). The mean percentage was 15·0 and the standard deviation 9·4, with a maximum value of 75·0 per cent. Values were highest in the central lowlands, south-west Scotland and in the islands, and lowest in the Southern Uplands, the Highlands and the north-east; absolute densities are greatest in the Lothians, the Clyde valley and lowland Ayrshire. This varies considerably with full-time, part-time and seasonal workers (Table 6); for whereas only 8 per cent of full-time workers were female, the percentages of part-time and seasonal workers were 66 and 42 respectively. There were also considerable variations with type of farm (Table 6); the proportions of full-time, part-time and seasonal female workers increasing with intensiveness, though the rank order remains the same.

Although no mappable data were available, information collected in the December census permits some analysis of the relative contribution of family and hired labour. Twenty per cent of male regular workers (ie full- and part-time) and 42 per cent of the women were members of the occupier's family, although there were considerable regional variations; in the Highlands the proportions were 28 and 72 respectively, rising to 63 and 92 per cent in Zetland, while in south-east Scotland the proportions were only 12 and 21 per cent. As with other differences in the labour force, these variations are partly a reflection of size of farm, for proportions will tend to be high where farms are small; on part-time and spare-time farms 46 per cent of all male workers and 77 per cent of all female workers were family workers, whereas on cropping farms,

which are characteristically large and have large labour forces, the proportions were only 12 per cent. The percentage is also low on hill sheep farms, possibly because many are run in conjunction with lowland farms and worked with hired labour.

In addition to the agricultural workers enumerated in the agricultural returns, the agricultural work force also includes, to a varying degree, the occupiers of agricultural holdings. Consideration of their distribution is deferred to Section 6, where maps of part-time holdings are discussed, such holdings being defined as those in which the stock and crops enumerated in the June census require fewer than 250 standard man-days (smd) of work; but it will be appropriate to conclude this section on agricultural labour with a brief consideration of the status of occupiers and their relation to employed labour.

According to the 10 per cent sample in the 1966 census, there were some 44,050 farmers, market gardeners and farm managers in Scotland, together with 3,650 crofters, compared with some 62,670 agricultural workers. Of course, 'occupier' is not synonymous with 'farmer', nor can agricultural workers in the agricultural census be equated with those in the population census, for the latter is concerned with those who consider themselves primarily farmers and farm-workers, and the former with anyone who occupies or works on farmland. Of the 56,800 holdings in Scotland in 1965, only 25,200 were estimated to provide full-time employment for one man, a large proportion of the remainder being crofts. The distinction between full- and

TABLE 6

Female Workers and Type of Farm in 1965

	Hill sheep	Upland	Rearing with arable	Arable rearing and feeding	Cropping	Dairy	Intensive	All
Percentage of female workers								
Full-time	3	6	6	5	4	11	19	8
Part-time	37	42	52	55	67	76	74	66
Seasonal	15	18	21	27	61	37	66	42
All	5	10	10	9	15	20	33	15

Source: *Scottish Agricultural Economics*, Vol 17

part-time holdings is, of course, somewhat arbitrary and their occupiers are not synonymous with full- and part-time farmers; it is recognised that there is a grey area on either side of the cut-off point where the criteria used are not very discriminating. A sample inquiry in 1965-7 revealed that some 800 holdings with more than 250smd were worked part-time while 2,200 holdings with less than the minimum for a full-time holding were in fact worked full-time. Of the 32,800 holdings that were nominally part-time, 25,300 provided less than 100smd of work, and some 16,000 of these have been classified as statistically insignificant and have since been excluded from the census.

On the basis of a sample survey in 1968/9, some estimates have been made of the number of occupiers who have other employment. These suggest that some 17,500 (or 39 per cent of all occupiers) have other jobs, a figure that has not greatly changed since a similar inquiry in 1959/60. Forty per cent of these occupiers were on nominally full-time holdings, 12 per cent of them occupying holdings with over 1,200smd, though the proportion of all occupiers with other jobs fell with increasing size of holdings. For 15,000 of these 17,500 occupiers, agriculture was a subsidiary occupation and, although the proportion declined with increasing farm size, this was also true of a majority of those with full-time holdings and even those with holdings of more than 1,200smd. Indeed, two-thirds of those with another job had acquired their non-agricultural job first, though this was true of only a third of those who occupied full-time holdings.

The survey also showed that occupiers were on average older than farm-workers, with 20 per cent aged 65 and over compared with 3 per cent of farm-workers. Female occupiers accounted for 12 per cent of all occupiers, although the proportion was higher on holdings with 100smd or less. The survey also showed that a quarter of those with another job were employed in agriculture.

These comments make clear that the dividing lines between farmers and non-farmers and between farmers and farmworkers are by no means well defined. Not only do farmers devote some time to work on their holdings, but some workers are also occupiers of other holdings; furthermore, some occupiers and some workers are only marginally involved in farming. The maps must accordingly be interpreted with care.

MACHINERY

Information about machinery is collected triennially at a special machinery census undertaken in February, but it is available only by counties and cannot be directly related to the June census. The maps of machinery in this atlas are thus included as a token of the much larger number that might be produced if suitable data were available at parish level; for information is collected on five kinds of tractor, stationary petrol and electric engines, various kinds of transport and load-hauling equipment, implements for tillage and cultivation, for sowing and for distributing fertiliser, for harvesting hay, silage and roots, for drying, storing and preparing crops, for dairying, and for a variety of other purposes. In 1962 such equipment was estimated to be worth £63 million, 30 per cent of which was on dairy farms and 22 per cent on cropping farms. Machinery and fuel represent 19 per cent of annual expenditure on Scottish farms, a percentage that varies somewhat with type of farm, with the lowest proportion on hill sheep farms; it also tends to decline with increasing size of farm, here measured by standard labour requirements (Table 7).

The complexity of the data to be mapped, as illustrated by the range of types of tractor and combine harvester, and the deficiencies of the county as a spatial unit are severe handicaps to the effective mapping of these data. Tractors are the most widespread implement and numbered 60,940 in February 1967, of which, 1,003 were under 10hp and a further 1,040 were tracked vehicles; after the rapid expansion during and after World War II, numbers are now fairly stable, though changes in composition have been taking place. Fig 22 shows the distribution of tractors (unweighted by type) for every 1,000 acres of crops and grass. The highest densities were in the Highlands and north-east, and the lowest in the Tweed valley. Interpretation is complicated by a number of factors: the large proportion of rough grazing in upland counties, the varying role that tractors play on the farm,

TABLE 7

Machinery Costs in 1965

Type of farm

smd	Hill sheep	Upland	Rearing with arable	Arable rearing and feeding	Cropping	Dairying
		Percentage of inputs (excluding seed and feeding stuffs)				
275–599	16	30	34	32	36	40
600–1,199	14	26	28	31	29	31
1,200+	17	24	23	24	26	27
All	16	26	26	28	27	28

Source: *Scottish Agricultural Economics*, Vol 17

the different mixes of large and small tractors, and the tendency for occupiers to keep tractors beyond the end of their economic life as stand-by equipment. In Zetland, for example, tractors under 10hp accounted for 23 per cent of all tractors and only 30 per cent were diesel tractors, whereas the proportions in Berwickshire were under 1 and 90 per cent respectively. None the less, tractors are the most widespread of major pieces of equipment and are more widely distributed than the intensity of agriculture might suggest; thus the Highlands, which contributed about 8 per cent of gross output, had 12 per cent of the tractors, and the North East 25 and 31 per cent respectively; for tractors are not only used to pull implements but also as transport for people and goods.

In 1967 binders still outnumbered combine harvesters by almost three to one, totalling 17,936 (compared with 28,893 in 1946) and 5,258 (compared with 189 in 1946) respectively. Nearly half the binders were in the North East, and ratios were particularly high on Orkney and Zetland, and lowest in south-east Scotland (Fig 23). Almost half the combine harvesters and more than half the larger self-propelled machines were in the East Central and South East Regions, the principal cereal growing areas (Fig 24).

Livestock farming has been much less mechanised than crop farming and, apart from shearing machines and the highly mechanised equipment of the broiler industry, mechanisation has been largely confined to dairying. Milking

machines, of which there were 36,329 in 1967, were most numerous in the South West, which contained 61 per cent of all units; but they were relatively most important in those lowland counties where dairying is of minor importance (Fig. 25). This inverse relationship is not an uncommon feature of machinery distribution, with a machine being relatively most important where it is least numerous.

MARKETS

Except in subsistence agriculture, the main objective of farmers is the production of commodities for sale, whether to other farmers or to those outside agriculture. Little is known in detail about the ways in which crops and livestock are moved off farms and about the routes they follow to the ultimate consumer. Some products, such as store stock, may be sold to other farmers and may move directly between farms; alternatively they may pass through markets, while other commodities may be sold through middle men or direct to processors, often on contract between producer and manufacturer. Some relationships can be inferred by examining the distribution of markets, using this term in the widest sense, and the maps in this section provide a small sample of the many kinds of outlets.

Slaughter houses are mainly in public ownership, and large establishments account for by far the greater part of the throughput (82 per

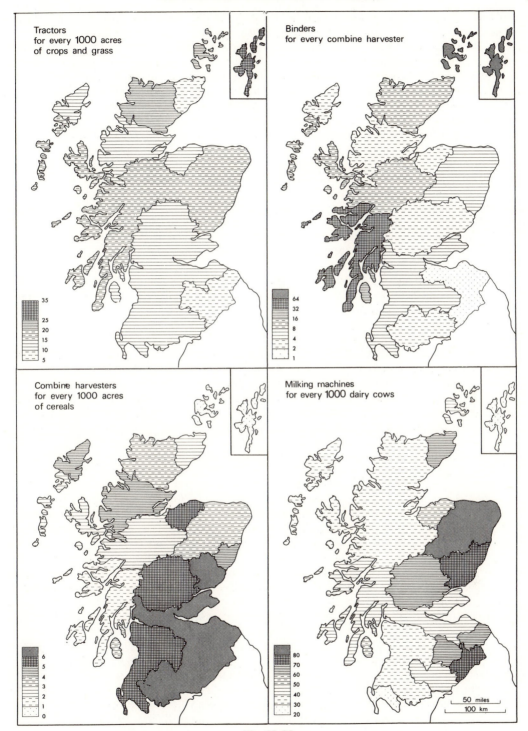

Figs 22–25

cent in 1962). Slaughter houses are located both in the main consuming areas and in the principal fattening areas of east and south Scotland (Fig 26), and, in the postwar period, there has been some tendency for the producing areas to acquire a larger share. In addition to those slaughter houses shown on Fig 26, there are also small establishments in the Highlands and islands which are used only occasionally. Scotland is not a major area for bacon production, and the term 'bacon factory' is increasingly a misnomer, for a variety of products is produced, especially in the larger plant; factories are located mainly in central Scotland, rather than in the main pig farming areas of the north-east (Fig 27). Distilleries show a rather different distribution (Fig 28). The largest concentration is in Banff and Moray, largely in the valley of the Spey and its tributaries, while others are scattered throughout the eastern margins of the Highlands and in Argyll. Most of these produce single malt whiskies, the character of which is believed to owe something to local supplies of peat and water, though most of the barley required for malting is now imported, whether from overseas or from England. The bulk of whisky is blended and distilleries producing grain whiskies are mainly found in central Scotland. Creameries are largely concentrated in the main dairying areas, particularly the coastal lowlands of Dumfriesshire, Kirkcudbright and Wigtown (Fig 29). Some of these act as collecting points for the onward transmission of milk in bulk to central Scotland, but this function is less important than in England, partly because distances are generally shorter, because a larger proportion is manufactured and because most of the milk from Scottish farms is collected by bulk tanker. Most creameries manufacture cheese, and some in Dumfriesshire produce condensed milk. Scattered creameries elsewhere on the islands and in the north-east are required to absorb seasonal surpluses, though actual quantities are small.

There are thus rather different pulls of market and producing area, and varying degrees of inertia in adjusting to changed sources of supply. Because of the dominant position of livestock and livestock products in Scottish farming and because livestock can be marketed at various stages in the production cycle, the great majority of movements off Scottish farms concern movements of livestock. For sheep, and to a lesser extent cattle and pigs, large numbers of stock are marketed more than once, first as store animals and then as fat stock, and the resulting patterns of marketing are rather different. According to a survey in 1963 some 2,751,000 store sheep and lambs passed through auction markets in that year, together with 117,000 store calves, 582,000 older store cattle and 251,000 store pigs, while an unknown, though large, number was handled by dealers outside the auction markets; direct sales between farmers were thought to be unimportant. Store sheep come mainly from the hill and upland farms, the former being of pure hill breeds, and the latter first-crosses with these breeds. Cattle come from a variety of sources. Like store sheep, some store cattle originate on hill and upland farms, but others come from lowland farms; another source of supply is dairy farms, both in southwest Scotland and in England, which provide surplus calves suitable for beef. Movements of pigs are probably more local.

Two sources of information have been used to map these aspects, the records of sales in the different markets and the results of an inquiry among inspectors in the Department of Agriculture and Fisheries for Scotland. The former is quantitative, but gives no indication of the source of the livestock marketed, whereas the latter is qualitative, but provides some impressionistic indication of movements. The information provided by the inspectors was obtained by questionnaire and by maps on which they were asked to record the main movements of stock and crops to and from their areas. It should, however, be appreciated that commodities which appear significant in a receiving area, providing much of its supply, may not appear so to the exporting area, accounting for only a small part of its sales; the reverse may also be true.

Store sheep markets are widely distributed throughout Scotland, many being located on the margins of upland and lowland, though there are smaller markets elsewhere in the Highlands and islands (Fig 30). Store cattle markets show a greater range of sizes, with the most important in north-west and central Scotland; many of

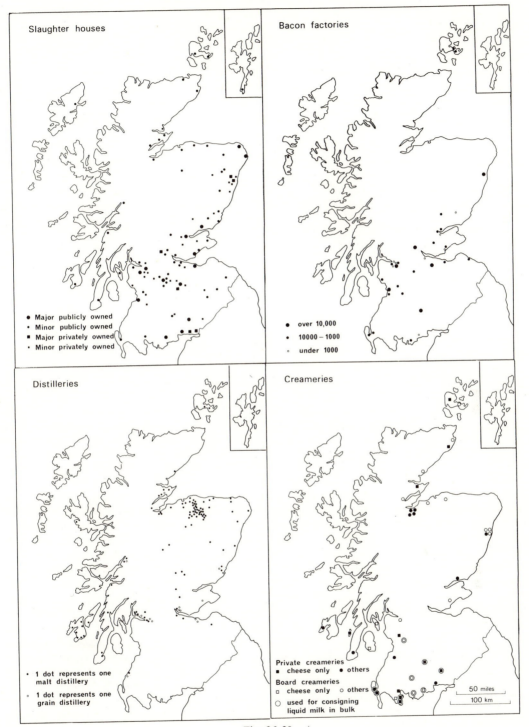

Figs 26–29

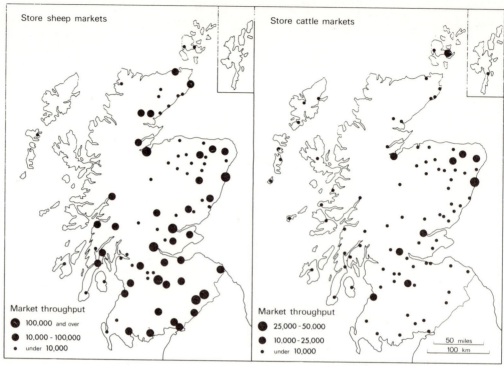

Figs 30–31

those in the Highlands are highly seasonal (Fig 31).

Four sample maps are included to show the wide range of movements that exist, though no indication is given of the scale of movement. Wintering of hill ewe replacements, less important than formerly owing to the cost of transport and of winter keep, is unusual in that the movement is two-way, the autumn migration to low-ground farms being reversed in spring when the ewe hoggs return to the upland pastures on which they were born (Fig 32). Movements of store sheep are broadly similar to the autumn flow, but, like those of store cattle, are one-way. The map of store cattle is more complicated, comprising not only movements from hill to lowland, but also those from south-west Scotland and from England (Fig 33). Movements of dairy herd replacements are on a small scale and chiefly involve the sending of replacements from the main dairy areas to more northerly counties, notably Orkney and Shetland and the north-east

(Fig 34). Movements of hay and fodder crops have been included to exemplify those movements which seem unimportant to the sender, but important to the receiver. The main movements are of hay from Stirlingshire and of hay and straw from the north-east to the Highlands and Islands; there are also movements from the Lothians and the Merse to the south-west (Fig 35).

Little is known about the costs of such movements, except that they are relatively high to and from the islands and, to a lesser extent, from the crofting counties generally; in a sample inquiry in 1961/2 both transport costs per mile and distances travelled were greater, as well as total costs, though the sample was a small one. Outside the crofting counties there was little difference between regions. Whatever the significance in terms of cost, these invisible links between farms and between farms and markets need to be borne in mind in interpreting the maps in this atlas.

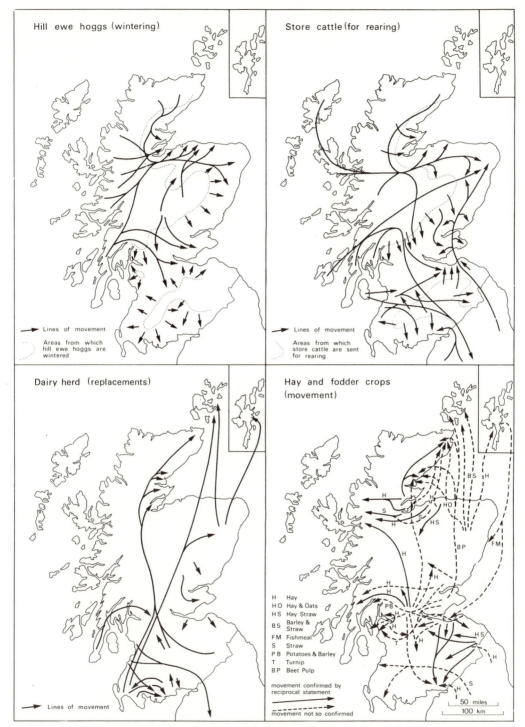

Hill ewe hoggs (wintering)

→ Lines of movement

⌇ Areas from which
hill ewe hoggs are
wintered

Store cattle (for rearing)

→ Lines of movement

⌇ Areas from which
store cattle are sent
for rearing

Dairy herd (replacements)

→ Lines of movement

Hay and fodder crops
(movement)

H Hay
HO Hay & Oats
HS Hay Straw
BS Barley & Straw
FM Fishmeal
S Straw
PB Potatoes & Barley
T Turnip
BP Beet Pulp

—→ movement confirmed by
reciprocal statement
--→ movement not so confirmed

50 miles
100 km

Figs 32–35

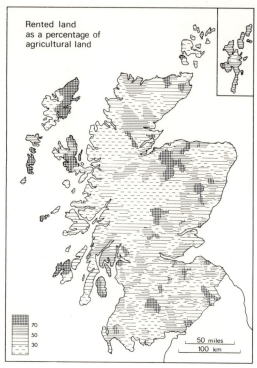

Rented land
as a percentage of
agricultural land

70
50
30

50 miles
100 km

Fig 36

TENURE

The term 'occupier' is neutral, implying nothing about the way in which the land is held. An occupier may own the land he farms or rent it from one or more owners and, in the seven crofting counties, he may be a crofter. How he farms the land will in part depend on his status and, if he is a tenant, on the status of his landlord; for example, there is certainly evidence that investment on occupied farms is greater than on tenanted farms, and it is probable that the owner-occupier has greater flexibility in his choice of farming systems.

In the nineteenth century, most farmers were tenants, often of large estates, though there would generally be one or more home farms and the woodlands and shootings would also be kept in hand. During this century, and particularly since World War I, many estates have been divided among their constituent farms and the proportion of land occupied by owners has

increased considerably until it had reached 59 per cent in 1969. The concept of rented land is not, however, a simple one, for although the proportion has changed little since 1962, there has almost certainly been a further move to owner-occupation which has been obscured by a tendency for farmers to transfer nominal ownership to legally separate persons or firms while retaining effective control. Fig 36 shows the distribution of rented land as a percentage of all agricultural land in 1960, the last year for which parish data are available. It exhibits no clear pattern, with high and low values juxtaposed in adjacent parishes; values were lowest in the Highlands and parts of the Southern Uplands and highest in the islands. These differences are in part due to the large number of crofts, regarded as tenanted for this purpose, among the holdings under 10 acres and the presence in the largest size group of large home farms and of farms being kept in hand by owners of large estates.

Apart from crofts, there are other kinds of tenancy that may affect the way in which land is farmed. The Secretary of State for Scotland holds some 420,000 acres (175,000ha) of land settlement estates, acquired since 1886 for a variety of purposes and let in smallholdings. Some 375,000 acres of these lie in the Highlands and are mainly crofts, but the remaining 44,000 acres are in the lowlands. Many of these holdings are no longer capable of providing an acceptable standard of living and it is government policy to create more viable holdings by amalgamation when this is possible; but this is a slow process owing to the high degree of security enjoyed by such tenants. The Forestry Commission also holds large acreages of agricultural land, most of which is tenanted. Much of this land is managed by the Department of Agriculture on the Commission's behalf, and is either land which is to be planted in due course or land which is too good to plant and which has been acquired as part of an estate. It is policy to sell the land which is not required for forestry, but difficult to do so where the land is tenanted. Land held by other public authorities, although let for agricultural use, is often subject to some restraint, as on land owned by the Ministry of Defence and used for military training, or held

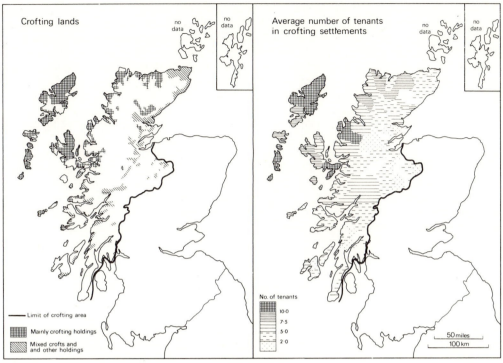

Figs 37–38

as gathering grounds by water authorities. Recreation may also be expected to impose restraints, and does so to some degree on that farmland, mainly rough grazings, held by the National Trust for Scotland. On privately owned land which is also used for field sports and rough grazing there may be restraints on agricultural use in the interests of sport, particularly on the 3 million or so acres used as grouse moors.

Crofting Tenure

A croft is a statutory agricultural smallholding in one of the seven crofting counties (Argyll, Caithness, Inverness, Orkney, Ross and Cromarty, Sutherland and Zetland) which is farmed by a registered crofter whose holding at the time of registration was either rented at a rent of not more than £50 per year or had an area of crops and grass not exceeding 75 acres (30ha). The crofter enjoys virtually complete security of tenure, but differs from other agricultural tenants in that he provides his own house and agricultural buildings; most crofts also have access to common rough grazings (see Fig 49). Roughly two-thirds of the crofts, which represent some 15,540 units are on the islands, a third of them in the Outer Hebrides; as Fig 37 shows, their distribution is predominantly coastal. Crofts are also important in Zetland, for which no comparable data were available and which contains 15 per cent of the crofts. Crofting communities are also appreciably larger in the islands and in the western mainland than they are in the eastern mainland, where many settlements consist of a single croft (Fig. 38).

Crofting legislation has maintained a system of smallholding which has disappeared elsewhere in the Highlands and in Scotland, but it has not maintained a viable agriculture. Many crofts are neglected by elderly or absentee tenants and, according to a survey in 1960, as many as 10 per cent of crofts in the Outer Hebrides were held by absentees. The 1961 Crofters Act facilitated subletting but in 1970 it was estimated that only 3 per cent of crofts provided a full-time occupation for the crofter, and most of these were in the eastern mainland; 85 per cent, on the other hand, had less than 100 smd. Furthermore, although crofts account for 14 per cent of the agricultural area of Scotland, they produce only 4 per cent of the agricultural output. Crofts are largely devoted to the production of store cattle and sheep and generally occupy poor land, though the finishing of stock on land improved with the aid of grants is being encouraged. In view of the poor quality of much of the land, the great security of tenure enjoyed by the crofters, their high average age and the small size of most crofts, there would be little prospect of converting all croft land into viable agricultural units; nor would the Crofters Commission, which has a general supervisory role, wish to do so, since it would result in the further depopulation of the crofting areas. Instead, attempts are being made to increase other sources of income, particularly from tourism.

SIZE OF HOLDING

The size of a farm can be measured in a variety of ways. Historically, in the absence of any alternative measure, most analyses have used the number of holdings in each size group as measured by the acreage of crops and grass, but data are now available both on total acreage of agricultural land and on the labour requirements as measured by standard man-days (smd). Yet, whatever the basis of measurement, the importance of different size groups can be assessed either by comparing the numbers of holdings in each class or by weighting these according to acreage or labour requirements. Each parameter measures a different aspect of farm size. Numbers of holdings focus attention on the occupiers and the land resources they command, acreages emphasise the areal extent and the contribution of the holdings to the landscape, while size of business shows the intensiveness of farming; none is satisfactory on its own. Two approaches, employing total acreage and standard man-days, have been selected to display the information on farm size for the parishes of Scotland; two reasons for doing so are that grouping by acreages of crops and grass alone can give a misleading impression in a country where rough grazing accounts for three-quarters of all agricultural land, while the large number of very small holdings tends to obscure the influence of larger holdings when the proportion of holdings in different size groups is calculated. Table 8 shows the difference between the average size of full-time and all holdings in 1969 on the basis of total acreage and crops and grass acreage in each of the major regions.

Table 9 shows the relationship between farms classified according to their size of business and their share of the farms acreage, and emphasises the misleading impression, in terms both of area occupied and share of standard labour requirements, which is given by the proportion of holdings.

TABLE 8

Average Size of Holdings in 1969

	Highland acres	North East acres	East Central acres	South East acres	South West acres	Scotland acres
			Crops and Grass			
Full-time	117	152	197	266	156	170
All holdings	20	84	135	173	105	78
			Total acreage			
Full-time	2,319	277	536	571	391	575
All holdings	355	163	371	365	258	279

Source: Agricultural Census 1969

TABLE 9

Farm Businesses, Acreages and Number of Holdings in 1968

smd	Percentage of holdings	Percentage of crops and grass	Percentage of total acreage	Percentage of smd
−250	59	12	16	7
250–599	17	18	20	16
600–1,199	14	28	27	28
1,200+	10	43	37	49
All	100	100	100	100

Source: Agricultural Statistics 1968

TOTAL ACREAGE SIZE GROUPS

Four acreage size groups have been selected for mapping, each map showing the proportion of agricultural land on holdings in the appropriate size group in 1969. The four groups are: under 100 acres (40ha); 100–299¾ acres (120ha); 300–999¾ acres (400ha); and 1,000 acres and over. These groups can be categorised very broadly as follows: holdings of under 100 acres are mainly smallholdings, often run on a part-time basis, though they also include some intensively worked holdings; holdings of 100–299¾ acres include a large proportion of the family farms, providing full-time employment but relying mainly on family labour; holdings of between 300 and 1,000 acres generally represent large farms relying on hired labour; and those of 1,000 acres and over are primarily large hill farms.

Over most of Scotland less than 10 per cent of the acreage of agricultural land was in holdings of under 100 acres (Fig 39), but there are three major exceptions: the central lowlands, where the proportion was less than 25 per cent; Buchan where it mainly exceeded that value; and the islands, notably the Outer Hebrides, Orkney and Shetland, where there are large numbers of crofts. Were it not for the large size of the parishes in the north-west Highlands, most of the coastlands there would also have been shown in higher categories (cf Fig 37). It should also be noted that the effective acreage of crofts is larger than might seem from the size of individual holdings, in that crofters have access to common pasture (Fig 49).

The central lowlands and Buchan were also the main areas for the predominantly family farms of between 100 and 300 acres, with holdings in this size group accounting for over 50 per cent of the agricultural land in some parishes (Fig 40), the prevalence of such farms in the latter area reflecting the small livestock rearing farms characteristic of north-east Scotland. Coastal areas of south-west Scotland also had a fair proportion of land in holdings of this size group, but only Orkney and Shetland among the islands had large numbers of such holdings.

Holdings of between 300 and 1,000 acres were most characteristic of east and south-east Scotland, particularly the Merse, the Lothians and east Fife (Fig 41). These are predominantly areas of large arable farms, as is the coastal lowland of Perth and Angus, where this size of farm was nearly as important; but the coastal areas of south-west Scotland, where farms of this size were also numerous, are predominantly devoted to stock and dairy farms. Only in the Highlands and to a lesser extent in the islands and the Southern Uplands did holdings in this size group account for less than 10 per cent of all farmland.

Holdings of 1,000 acres and upwards are largely hill sheep farms which dominate the Highland parishes and, to a lesser extent, those of the islands and the Southern Uplands, accounting for 75 per cent and more of the agricultural land in most parishes (Fig 42); for hill-sheep farms averaged almost 4,000 acres (1,600ha) in 1969, compared with under 300 acres (120ha) for cropping farms. In contrast with the preceding maps, it is only in the lowlands that less than 10 per cent of agricultural land was in farms in this size group.

Although these distributions bear some relation to farm type and land quality, the relationship is not a simple one, for some of the smallest farms are located on some of the poorest land. A rather different picture would have emerged if the acreage under crops and grass had been used as the basis for classification; for, whereas the average size of full-time holdings in the Highlands in 1968 was 2,326 acres (962ha) of agricultural land, it was only 115 acres (48ha) of crops and grass (Table 8).

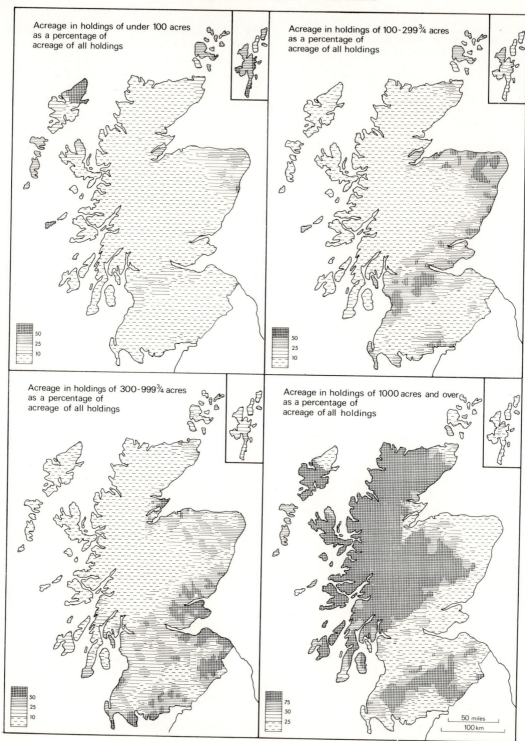

Acreage in holdings of under 100 acres as a percentage of acreage of all holdings

50
25
10

Acreage in holdings of 100-299¾ acres as a percentage of acreage of all holdings

50
25
10

Acreage in holdings of 300-999¾ acres as a percentage of acreage of all holdings

50
25
10

Acreage in holdings of 1000 acres and over as a percentage of acreage of all holdings

75
50
25

50 miles
100 km

Figs 39–42

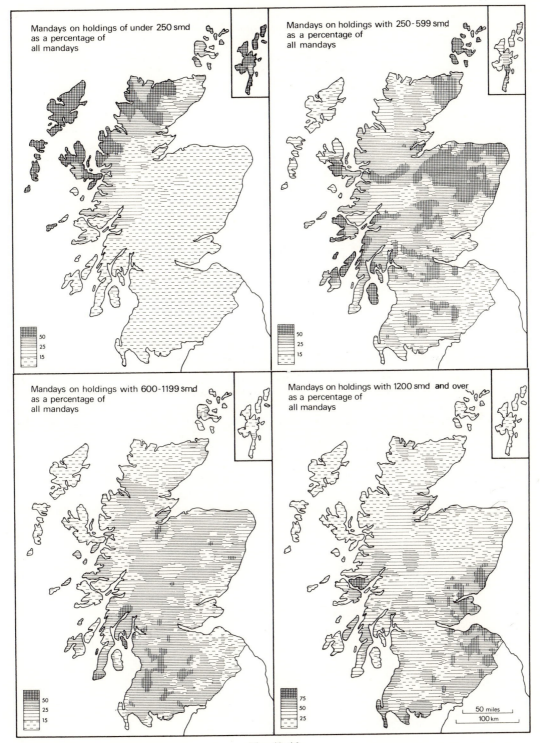

Figs 43–46

MAN-DAY SIZE GROUPS

As Table 9 shows, a major weakness of total acreage as an indicator of farm size is that all acres are treated as equal and no account is taken of the varying intensity with which land is farmed. Yet a hill sheep or upland farm, with a low carrying capacity, may be quite a small business as measured by its turnover, investment, sales or labour force. Unfortunately, no comprehensive data are collected on these parameters, so that the only possible approach is to convert the livestock numbers and crop acreages in the census into some measure of intensity by the use of standard factors. For some years, the agricultural ministries in the United Kingdom have been classifying farms on the basis of standard labour requirements and these are employed here as an alternative measure of farm size in order to correct the impression given by Figs 39–42.

Four size groups are recognised: under 250smd; 250–599smd; 600–1,199smd; and 1,200smd and over. Unlike the acreage classes, these groups can be identified unambiguously with different kinds of farm. Those with under 250smd are part-time holdings; the second group corresponds with farms providing work for 1–2 men which can be broadly equated with family farms; the third group comprises large farms with work for 3–5 men; and the fourth comprises very large farms providing work for 6 or more men. It must be recognised that these figures are based on standard factors, and so are unlikely to match labour actually used, and that the lowest cutoff point, 250smd, is only a rough and ready guide (cf p 29).

Fig 43, showing the proportion of all man-days on holdings of less than 250smd, ie part-time holdings, bears some resemblance to Fig 39 in that the highest values were to be found in the islands, notably the Outer Hebrides and Shetland; but there were also high values in the north-west mainland, rather than in Buchan, and the distribution more strongly resembles that of land in crofts (Fig 37). The map of family farms, ie those with between 250 and 600smd, shows that such holdings were widespread, but north-east Scotland, with its numerous stock rearing farms, was particularly prominent; only the Outer Hebrides and the arable farming districts of eastern Scotland had low values (Fig 44). Large farms, ie with between 600 and 1,200smd, were also widespread, with low values only in the north-west and south-east (Fig 45), whereas very large farms were most important in the main areas of arable farming in east and south-east Scotland (Fig 46). This last map is in marked contrast to the corresponding map of acreage size groups (Fig 42), though it should be noted that over most of the Southern Uplands and the Highlands south of the Great Glen more than 25 per cent of all man-days on holdings were in this category. The patterns shown by these maps are markedly more complex than the comparable maps of size groups by total acreage, largely because farms in the uplands, though covering large acreages, represent businesses of very varying sizes; 45 per cent of hill sheep farms had between 250 and 600smd in 1968, 34 per cent had 600 to 1,200 and 21 per cent had more than 1,200.

Farm sizes, whether measured by acreages or by standard man-days, have been increasing, though no satisfactory basis exists for comparison over any long period. For this reason, this topic is not discussed in the historical section, but there is no reason to suppose that the relative importance of holdings of different sizes throughout the country has changed greatly.

3

Land Use and Crops

LAND USE

In 1965, the area recorded in the agricultural census as under tillage (ie crops and fallow), grass and rough grazing was 16,624,993 acres (672,813ha); by 1972, partly owing to a reduction in the area of agricultural land, but primarily to changes in the administration of the census, whereby holdings of little agricultural significance were excluded, this figure had been reduced to 15,302,222 acres (619,281ha). Of these totals, almost three-quarters were rough grazing, and of the remainder, just over a third were devoted to tillage; but the boundaries between these categories are by no means clear cut, especially that between grass and rough grazing, which grade imperceptibly into each other. Similarly, at the extensive margin, there is no sharp dividing line between rough grazing and rough land not used for agricultural purposes.

ROUGH GRAZING

The dominant fact of the land-use map of Scotland is the large area under rough grazing, which occupies virtually all the Highlands and Southern Uplands and most of the islands (Fig 47). It reaches to sea level in the north-west, on poor soils and under a moist, windy climate, but more characteristically gives way to improved land at about the 800ft (240m) contour in the sheltered, drier valleys of the south-east, where cultivated land may extend above 1,000ft (300m) in places. The term 'rough grazing' covers a wide range of vegetation types, from bent-fescue pastures on the lower reaches of the hills in south Scotland, through molinia-dominated grass moors in the uplands of south-west Scotland and the heather moors of the eastern Grampians, to sedge moors on deep peat in the north-west. There are corresponding differences in carrying capacity, which may range from a ewe to the acre in the south to ten or more acres to the ewe in north Scotland; but measurement of stocking is difficult because of inadequate knowledge about the extent of land actually grazed. The acreage recorded as rough grazing now includes the whole acreage of deer forests, whether this is grazed or not, although, when only those areas grazed by sheep were recorded in the past, there were nearly a million acres (400,000ha) which were regarded as not grazed (although there is little doubt that some confusion existed in occupiers' minds and that some double counting occurred); this change, together with the elimination of some errors, increased the area returned as rough grazing by nearly 1½ million acres. It is doubtful if land above 2,000ft (600m) contributed very much to the carrying capacity of the Scottish moorlands and the type of farming map produced by the Department of Agriculture and published in 1944 showed large areas in the central and northern Highlands as of little agricultural significance; there has since been evidence of further decline in effective agricultural use of some of the poorer rough grazing (though the agricultural census, which records the land nominally in farms, cannot reveal this). At the same time, there have been attempts to upgrade the better rough grazings to grass pasture, although how much has been improved is unknown; a war-time survey suggested that only about 250,000 acres (100,000ha) was suitable for improvement, but it is now thought that there are economically viable ways of improving a larger acreage. Even so, by far the greater part of this land, which suffers from poor soils and a harsh climate, will remain as rough grazing if

it remains in agricultural use at all.

The proportion of rough grazing, both on farms and within parishes, varies as widely as does its carrying capacity and ranges from over 90 per cent in most parishes in the Highlands and islands, to under 10 per cent in those of the lowlands of the east and north-east (Fig 48), though even in the lowlands there are patches of rough land and some extensive areas of lowland bog. As will be seen in the discussion on types of farm (Chapter 6), the proportion of rough grazing varies greatly with type of farm and is in fact a diagnostic feature of hill sheep farms. Some 1,250,125 acres (505,926ha) of rough grazing are common pastures over which those in the crofting townships have rights to graze stock. Proportions are highest in the Outer Hebrides and Shetland where more than half all rough grazing is common grazing (Fig 49). Some former common grazings have been apportioned among individual crofters and this change was encouraged as a means of facilitating land improvement; altogether about 40,000 acres (16,200ha) of common pasture were upgraded by 1971.

A large part of the rough grazings is used for field sports, especially deer stalking and grouse shooting. Deer forests occupy much of the poorer land, including that not grazed by sheep, and deer are estimated to roam over some 7 million acres (2·8 million ha), conflicting with farming mainly during hard winters when they may invade farmland in search of food. Much of the 3–4 million acres of heather moor are used as grouse moors, which are burnt regularly to ensure a supply of young heather for grouse; they are also grazed by sheep and it has been suggested that stocking would not be greatly increased if shooting stopped. For the past 150 years, sheep have been the principal and often the only livestock grazing the uplands, though cattle (once a major component of hill livestock) have been increasing with government encouragement in the postwar period. Some authorities believe that long-continued sheep grazing of semi-natural vegetation, in conjunction with burning (which is virtually the only management tool in hill grazing, apart from control of stocking), has led to a deterioration of soil and vegetation, especially in the wetter

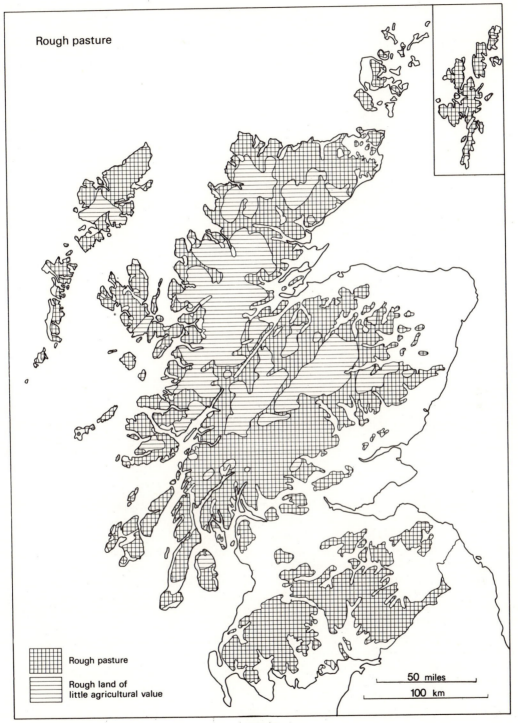

Rough pasture

▦ Rough pasture

▤ Rough land of
little agricultural value

50 miles

100 km

Fig 47

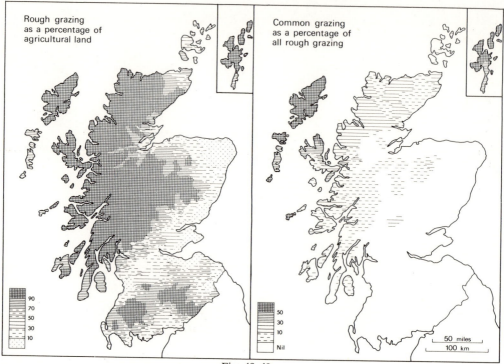

Figs 48–49

west; but this view is disputed by agriculturalists. Similarly, the spread of bracken, now occupying perhaps half a million acres (cf 200,000ha) of the rough grazings, is thought to be due, at least in part, to the replacement of mixed grazing by sheep and cattle with exclusive grazing by sheep.

CROPS AND GRASS

Although rough grazings occupy most of the area used for agriculture, they contribute only a small part of agricultural output; thus, hill sheep farms, which account for two-thirds of the rough grazings, are estimated to produce only 6 per cent of total output. Of course, this understates their contribution, since the stock bred and reared on the uplands plays a vital part in lowland agriculture and the main products of the rough grazings (store sheep, store cattle and breeding stock) are sold to Scottish farms and so do not enter into assessments of Scottish agricultural output. Nevertheless, the quarter of the agricultural land which is devoted to grass

and to other crops is of far greater importance. These accounted for 4,304,868 acres (1,742,180ha) in 1965, and for 4,163,409 acres (1,684,932ha) in 1972. Of course, Fig 50, showing the proportion of agricultural land in crops and grass, is the mirror image of Fig 48, but, despite the deficiencies of the parish as a mapping unit, it defines quite clearly those areas where the bulk of Scottish agricultural production takes place; a detailed land-use map would, of course, show that, as in the uplands, there are many pockets of land devoted to other uses. North-east Scotland provided the largest block of land with three-quarters of agricultural land under crops and grass, and most of the remainder was in parishes along the east coast; only the lowlands of Ayrshire and Dumfriesshire did not conform to this pattern.

TILLAGE

Tillage, or land under fallow or crops other than grass, is the least ambiguous of all the major

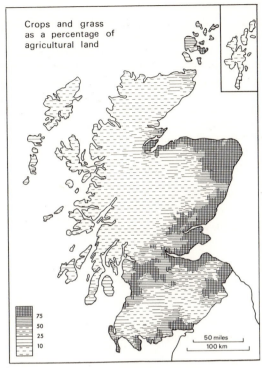

Crops and grass as a percentage of agricultural land

75
50
25
10

50 miles
100 km

Fig 50

categories of land use recorded in the agricultural census, and although it represents less than a tenth of the agricultural land, the sale of crops provides a fifth of agricultural output and large quantities of arable produce are consumed by livestock on the farms on which it is grown (though the percentage of output attributed to crops has fallen from 23 per cent in 1965 to 19 in 1972). The acreage under tillage, too, had fallen from 1,519,706 acres (615,025ha) in 1965 to 1,449,333 acres (586,545ha) in 1972.

Land under tillage is devoted to a large number of crops, each with its own requirements and each playing a part in cropping systems. Because the different crops can be readily identified and are separately recorded in the agricultural censuses, there is a danger of giving too much prominence to crops in any account based upon the censuses and, for these reasons, the text describing the various crops has been deliberately kept short. It will therefore be useful, by way of introduction, to survey briefly the principal crops on a comparable basis, with

particular reference to their regional distribution, the proportion of holdings on which each is grown and the types of farming with which they were associated.

DISTRIBUTION OF THE PRINCIPAL CROPS

While the distributions of the individual crops are shown on the succeeding maps, they cannot readily be compared, and Table 10 has therefore been prepared to show their distribution in 1965 among the five regions of the Department of Agriculture and Fisheries for Scotland; for comparative purposes, and to show the degree of stability over this seven-year period, selected figures for 1972 are also given. The proportion of the tillage acreage in each region provides a reference point for the individual crops; if these were of equal importance they should occupy the same proportions, though it should be noted that the regions vary in size, as do the acreages devoted to tillage. In general, the East Central and South East Regions had above average shares of the principal cash crops, viz wheat, barley, potatoes, sugar beet and vegetables, but below average proportions of the main fodder crops. These regions are, of course, only very crude indicators of location, but even this analysis has shown that there are marked variations in the degree of localisation, crops such as small fruit and sugar beet (no longer recorded in 1972, but with 78 per cent of the acreage in the East Central Region in 1965) being highly localised and others, such as barley, being more widely dispersed. There were some differences between the percentages in 1965 and those in 1972, but the ranking of regions remains much the same; the most notable feature is the further concentration of tillage and of many of the tillage crops in the East Central and South East Regions at the expense of other regions.

PROPORTIONS OF HOLDINGS WITH EACH CROP

The proportion of all holdings on which each crop is grown also varied widely. Table 11 records the proportion of holdings in each region on which each of the major crops was grown in 1967; it should be noted that these estimates were made before the administrative re-organisa-

TABLE 10

Principal Crops in 1965 and 1972

	High-land	North East	East Central	South East	South West	Scot-land
Percentage of tillage in each region						
1965	7	33	27	18	15	100
1972	6	32	29	20	13	100
Percentage of wheat in each region						
1965	5	13	42	33	6	100
1972	5	18	34	38	5	100
Percentage of barley in each region						
1965	4	32	28	23	12	100
1972	5	31	29	20	14	100
Percentage of potatoes in each region						
1965	6	18	48	14	14	100
1972	5	17	54	14	9	100
Percentage of vegetables in each region						
1965	1	16	38	33	11	100
1972	1	13	39	39	7	100
Percentage of small fruit in each region						
1965	1	3	82	6	7	100
1972	3	7	82	4	5	100
Percentage of oats in each region						
1965	10	44	15	11	20	100
1972	10	43	18	15	13	100
Percentage of turnips and swedes for stock feeding in each region						
1965	8	43	18	16	15	100
1972	9	43	20	18	11	100
Percentage of kale and cabbage for stock feeding in each region						
1965	13	12	26	14	35	100
1972	12	8	30	13	38	100
Percentage of rape in each region						
1965	17	6	22	24	31	100
1972	15	6	25	28	26	100

Source: Agricultural Censuses

tion of the census led to the elimination of many of the smaller holdings, though broadly similar results are obtained when the proportion of holdings with the crop is calculated in relation to all holdings with some tillage. In general, the fodder crops were not only more widely grown than cash crops, but also appeared on a far larger number of holdings. Potatoes were a notable exception, with small acreages being grown on a large number of holdings, contrasting markedly with wheat, although the acreage under both crops was approximately equal.

CROPPING AND TYPE OF FARM

Another important aspect of the cropping is the relationship of crop production to type of farm

TABLE 11

Proportion of Holdings with Each Crop in 1967

	High-land	North East	East Central	South East	South West	Scot-land
Percentage of holdings with tillage						
	63	82	80	70	60	70
Percentage of holdings with wheat						
	1	3	22	15	3	5
Percentage of holdings with barley						
	4	35	49	43	27	25
Percentage of holdings with potatoes						
	48	52	59	33	30	45
Percentage of holdings with vegetables						
	1	1	7	7	3	3
Percentage of holdings with small fruit						
	—	1	13	5	3	3
Percentage of holdings with oats						
	45	68	48	44	37	50
Percentage of holdings with turnips and swedes						
	19	60	47	42	30	37

Source: Agricultural Census 1967

TABLE 12

Principal Crops and Type of Farm in 1968

Hill sheep	Upland	Rearing with arable	Rearing with intensive livestock	Arable rearing and feeding	Cropping	Dairy	Intensive	Part- and spare-time	All holdings
			Percentage of acreage under tillage						
—	9	15	2	8	43	15	2	5	100
			Percentage of acreage under wheat						
—	—	7	2	8	70	11	2	1	100
			Percentage of acreage under barley						
—	4	14	3	8	48	18	2	3	100
			Percentage of acreage under potatoes						
—	2	5	1	5	65	14	2	5	100
			Percentage of acreage under oats						
1	19	21	2	10	23	13	—	12	100
		Percentage of acreage under turnips and swedes for stockfeeding							
1	16	24	2	10	27	14	—	6	100

Source: *Agricultural Statistics 1968*

(as defined in Chapter 6). Not surprisingly, the principal type of farming with which crops are associated is the cropping farm; in 1968, of the sixteen crops for which information was published, cropping farms were the leading type on nine, and for eight of these crops, such farms grew more than half the acreage recorded. Once again the major distinction was between cash crops and fodder crops, with the latter much more widely dispersed on a number of different types of farm. Table 12 shows the proportion of the Scottish acreage on each type of farm for five of the principal crops; the proportion of tillage is included as a reference point. The highest proportion of any field crop on a single type of farm in 1968 was sugar beet, with 90 per cent of the acreage on cropping farms.

Other types of farm were more important than cropping farms among the minor fodder crops. In 1968, dairy farms were the leading type for mixed corn (37 per cent), followed by upland (22 per cent) and part-time holdings (20 per cent). Dairy farms were also the leading type for kale (41 per cent) and upland farms for rape (35 per cent). Intensive farms were the leading type for soft fruit (55 per cent), followed by cropping farms (38 per cent), while for vegetables the ranking was reversed, with cropping farms accounting for 55 per cent and intensive farms for 35 per cent.

The main outlines of regional differences in cropping will now be clear and the topic will be further discussed at the end of this section, when crop rank (as measured, not by acreage, but by standard labour requirements) is examined and crop combinations identified.

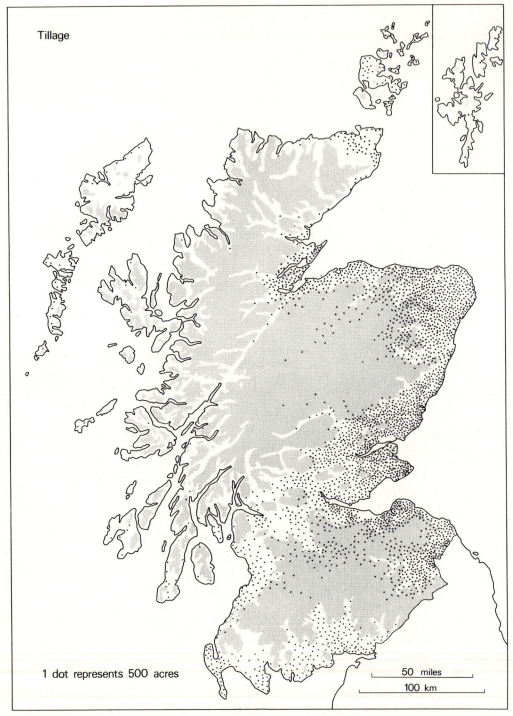

Tillage

1 dot represents 500 acres

50 miles
100 km

Fig 51

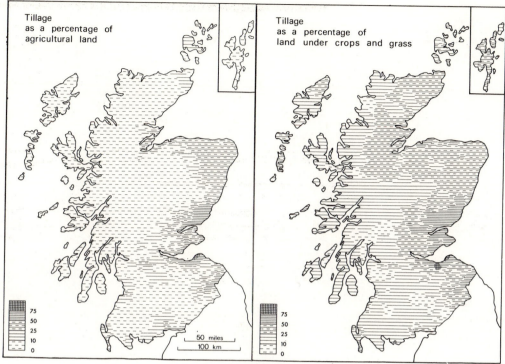

Figs 52–53

TILLAGE

Area: 1,519,706 acres (615,025ha) in 1965; 1,449,333 acres (586,545ha) in 1972

Percentage of agricultural land: Mean 18·5; SD 18·3; Max 78·5; Min 0·0

Percentage of land under crops and grass: Mean 30·7; SD 15·7; Max 80·3; Min 0·0

Fig 51 shows the distribution of land used for tillage in 1965; as would be expected, such land was to be found mainly in the eastern lowlands, from the Merse to the Moray Firth, in areas where friable soils derived from glacial drift and a relatively dry and warm climate provide the most favourable conditions for crop production in Scotland. Smaller acreages were to be found in the central lowlands and in south-west Scotland, but throughout the north-west and the islands (except Orkney) the amount of land in tillage is generally too small to be shown at this scale.

This map should be read in conjunction with Fig 52, which shows the proportion of agricultural land devoted to tillage and reveals a contrast between the lowlands north of Stonehaven and those further south; for whereas in the Merse, the Lothians, east Fife and the coastal lowlands of Perth and Angus more than half the agricultural land was in tillage, the proportion in Buchan and around the Moray Firth was between 25 and 50 per cent. Only in the central lowlands, the coastal lowlands of the south-west, the plain of Caithness and Orkney were percentages greater than 10.

Fig 53, recording the percentage of land in crops and grass devoted to tillage, is broadly similar, but shows that there were few areas where land in tillage accounted for less than 10 per cent of admittedly small areas of improved land even in the west and north (cf Fig 50).

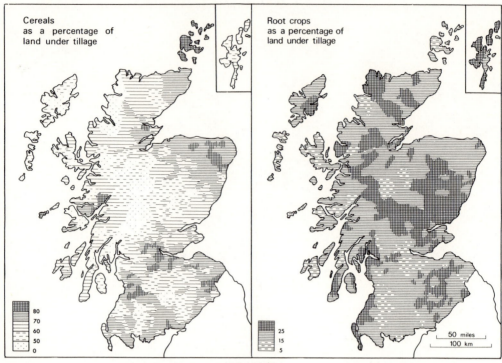

Figs 54–55

CEREALS

Area: 1,092,207 acres (442,016ha) in 1965; 1,143,481 acres (462,767ha) in 1972

Percentage of tillage area under cereals: 67·2; SD 12·1; Max 89·4; Min 0·0

Most of the acreage used for tillage is devoted to cereals, which accounted for 53 per cent of sales of crops in 1965 and 10 per cent of total sales (and 64 per cent and 10 per cent respectively in 1972). Fig 54, recording the percentage of land under tillage on which cereals were grown in 1965, shows that cereals accounted for 70 per cent or more in all those areas where there were large areas under tillage, with the exception of lowland Angus and Perth, where potato growing is important; indeed, in a few scattered areas, cereals occupied 80 per cent or more of the tillage, notably in Orkney. Only in parts of the Highlands and Islands did cereals occupy less than half the area used for tillage; for they are relatively easy crops to grow (especially where they can if necessary be harvested green, as with oats). They play a major part in rotations and have lent themselves readily to mechanisation.

ROOTS

Area: 339,563 acres (137,421ha) in 1965; 227,017 acres (91,974ha) in 1972

Percentage of tillage area under roots: Mean 21·6; SD 6·4; Max 52·5; Min 0·0

The corresponding map showing roots as a percentage of land under tillage is less clear, though it demonstrates that roots are rarely as important as cereals (Fig 55). Only in a few parts of Scotland did roots account for less than 15 per cent of the area under tillage but in few parts did they occupy more than 30 per cent. In large measure this reflects the importance of potatoes (not strictly roots) as both a cash and a domestic crop and of turnips as a fodder crop. Of the areas which are important for tillage crops, the lowlands of Perth and Angus stand out, with more than 25 per cent in roots. Most of the remaining high values correspond with areas where there was only a small acreage of tillage, where conditions are generally unfavourable to handling cereals and where a few potatoes may be grown for domestic use and some turnips for winter feed.

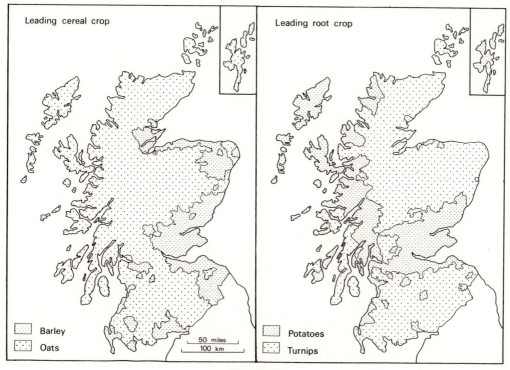

Figs 56–57

LEADING CEREAL

Three main cereals are grown in Scotland, in the descending rank order of barley (including bere), oats and wheat; a small acreage of mixed corn is also grown. Oats were long the major grain but were replaced by barley in 1965, the dominance of barley spreading gradually from the south-east; Fig 56 thus represents merely a stage in a process which has gone further since 1965. None the less, oats are often preferred in those areas where little land is devoted to tillage because of their greater tolerance of acid conditions, and because of the value of oat straw and the possibility of harvesting oats green if the weather does not permit, ie in the uplands and the islands. Barley, on the other hand, was the leading cereal throughout the lowlands, and neither wheat nor mixed grain was sufficiently important to displace it anywhere.

LEADING ROOT CROP

In 1965 there were three main root crops in Scotland, in the descending rank order of turnips and swedes (which are not separately distinguished in the census), potatoes and sugar beet; a small acreage of mangolds was also grown. The growing of sugar beet has since been discontinued following the closure of the British Sugar Corporation factory at Cupar, Fife, though it was nowhere sufficiently important to displace either turnips or potatoes as the leading root crop in 1965. Potatoes were the leading root crop throughout central Scotland, and were also important along the west coast, both in those parts of coastal Ayrshire that specialise in early potatoes and in the crofting districts of the north-west, though, as the discussion of the potato crop subsequently shows, potatoes are in fact three crops, seed, earlies and main crop (Fig 57). Turnips and swedes were the leading root crop both in the Highlands and Southern Uplands, where only a small acreage of root crops is grown, and also in the more important tillage areas outside the central lowlands, notably in the Merse.

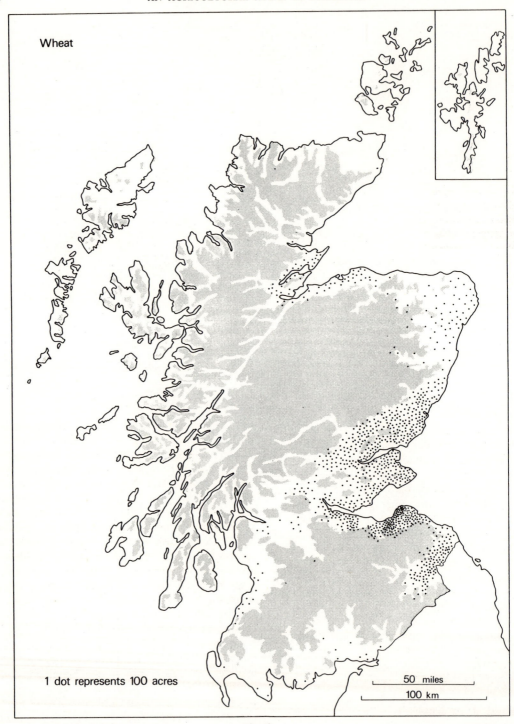

Wheat

1 dot represents 100 acres

50 miles

100 km

Fig 58

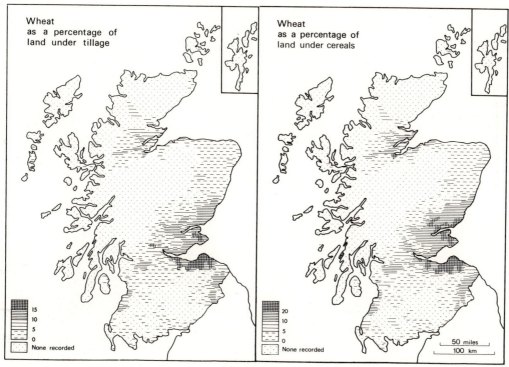

Figs 59–60

WHEAT

Area: 99,404 acres (40,299ha) in 1965; 81,437 acres (32,958ha) in 1972

Percentage of tillage under wheat: Mean 3·6; SD 5·1; Max 26·3; Min 0·0

Percentage of cereals under wheat: Mean 5·1; SD 7·2; Max 37·2; Min 0·0

Wheat, the third-ranking cereal by acreage but the second by contribution to output, accounted for 1·8 per cent of agricultural output in 1965, or 9·5 per cent of that from crops; wheat is used for both milling and stockfeed in roughly equal proportions, though these vary greatly from year to year, as does the crop acreage. It is generally regarded as the most difficult of the cereals to grow and was the least widely grown cereal in Scotland, where it reaches its northern economic limit (Fig 58). It was grown on relatively few farms and was largely confined to east and south-east Scotland, only a small acreage being grown in south-west Scotland or north of

Stonehaven; there was a small outlier in the lowlands around the Moray Firth, which are relatively sunny and dry, and very small acreages were grown near Wick and in Orkney. The late onset of spring and inadequate summer warmth are major handicaps, for the quality of grain suffers from a late harvest, and wheat, more than other cereals, requires good land. As Fig 58 shows and Fig 59 and 60 confirm, the principal area for wheat is in East Lothian, where it accounted for more than 15 per cent of the land under tillage in 1965 and more than 20 per cent of that under cereals. In the Merse wheat faces strong competition from barley and in the lowlands north of the Tay late harvests are likely to conflict with the demands of the potato harvest. Over the uplands, in the north-west and on most of the islands none is grown, and in most of the areas in north-east and south-west Scotland, where some wheat is grown, less than 5 per cent of either the tillage or the cereal acreage was devoted to the crop.

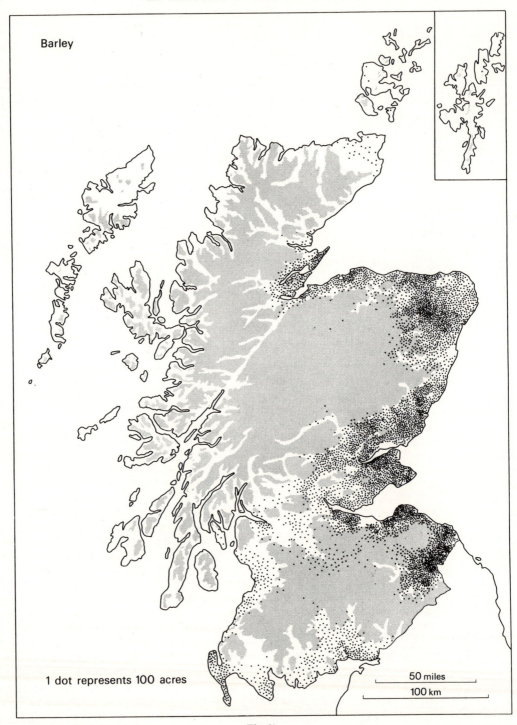

Barley

1 dot represents 100 acres

50 miles
100 km

Fig 61

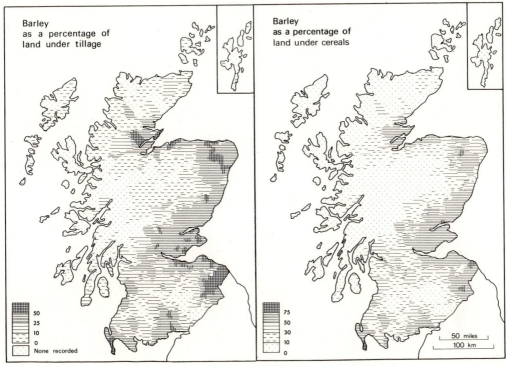

Figs 62–63

BARLEY

Area: 558,522 acres (226,034ha) in 1965;
820,987 acres (332,253ha) in 1972

Percentage of tillage under barley: Mean 24·4;
SD 18·5; Max 66·3; Min 0·0

Percentage of cereals under barley: Mean 34·4;
SD 24·9; Max 87·7; Min 0·0

Barley is the leading cereal in Scotland and the acreage devoted to it has increased greatly since 1960. It accounted for 6·4 per cent of agricultural output in 1965 and 34 per cent of that from crops. Barley was much more widely grown than wheat and by approximately five times as many farmers. None the less, its distribution in 1965 was predominantly eastern, in the lowlands from the Merse to the Moray Firth, and there is an appreciable number of northern and western parishes in which none was recorded in 1965 (Fig 61). Barley is grown for seed, for malting and for feeding, and about a third is retained on the farm on which it is grown for feeding to livestock. It is susceptible to soil acidity but has been able to extend its range by virtue of the subsidised liming that has been widely practised since 1939. The crop prefers warm, dry conditions, but spring varieties can be sown quite late, an advantage in a climate where late harvests may delay the sowing of cereals in autumn, and the short-strawed varieties bred in the post-war period can stand up better to wet conditions; higher yields also give it an advantage over oats, provided the crop can be combined. The principal area, judged by the proportion of the tillage acreage in 1965, was the Merse, with over 50 per cent under the crop; but most other cropping areas had at least 25 per cent, as did coastal areas in the south-west, though the acreage grown there was not large (Fig 62). The crop was also popular in Buchan, where it accounted for more than 25 per cent of the tillage acreage and 50 per cent or more of that under cereals (Fig 63). All the principal areas for crops had 50 per cent or more of the cereal acreage devoted to the crop. Large farms are often well equipped to harvest barley and to dry and store it, advantages which the small farmer of the north-west usually lacks.

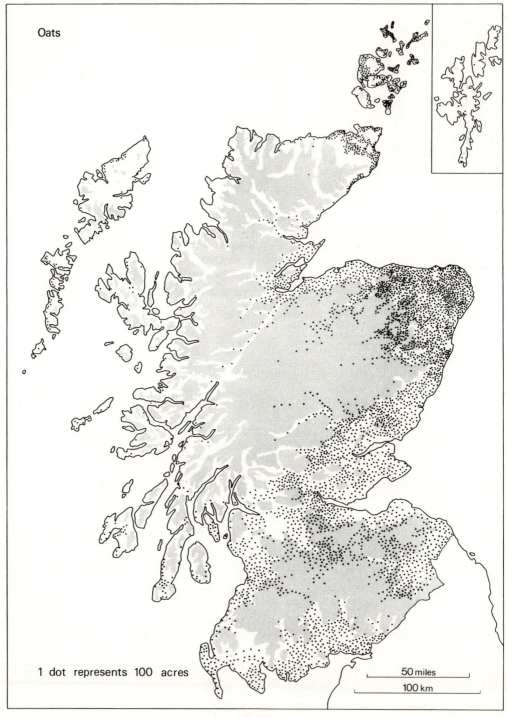

Oats

1 dot represents 100 acres

50 miles

100 km

Fig 64

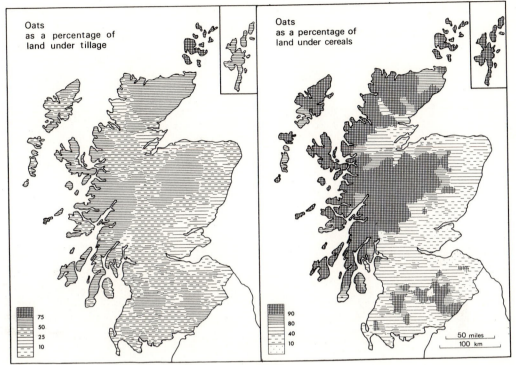

Figs 65–66

OATS

Area: 432,872 acres (175,183ha) in 1965; 232,978 acres (94,286ha) in 1972

Percentage of tillage under oats: Mean 38·9; SD 19·1; Max 88·1; Min 0·0

Percentage of cereals under oats; Mean 59·8; SD 30·1; Max 100·0; Min 0·0

Oats were the second ranking cereal, though they have been losing ground steadily to barley; they accounted for 1·6 per cent of agricultural output in 1965 and for 8·7 per cent of that from crops. Rather more than half the crop is retained on the farms on which it is grown for stockfeeding, and four-fifths goes for feeding to livestock. Oats are the most widespread cereal, though their distribution in 1965 was still primarily eastern, especially in Buchan, where oats play an important role in rotations (Fig 64). In southern Scotland their distribution was much more uniform than that of any other cereal. Oats are more tolerant of soil acidity than other cereals and they can be harvested green; they are also more suitable than barley for harvesting by binder. Many farms in the

west and north have been too small to invest in combines (even assuming their farms are suitable), but this is changing and with it the further replacement of oats by barley. Oats were also grown by more than twice as many farmers, though the acreage per farm was much smaller. To some extent, the crop is a residual legatee, being grown in areas unsuitable for other cereals, a view which is confirmed by Figs 65 and 66; for it was in western and northern parishes that oats were relatively most important. In most of the western half of Scotland 50 to 75 per cent or more of the tillage acreage was devoted to oats (though the acreages grown would often have been small); and in most of the uplands, the north-west and the islands, 90 per cent or more of the cereal acreage was under oats.

MIXED CORN

In 1965 there were 2,483 acres (1,005ha) under mixed corn, a figure which had increased to 8,080 acres (3,270ha) by 1972. Its distribution is widespread but patchy and in 1965 it was not grown in the great majority of parishes.

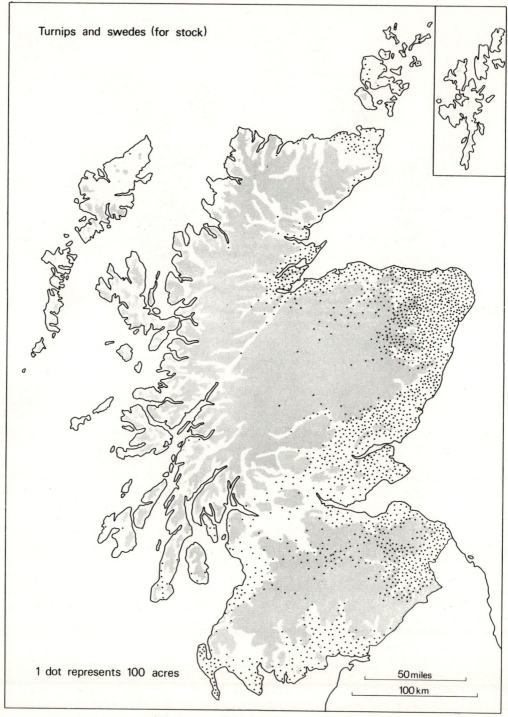

Turnips and swedes (for stock)

1 dot represents 100 acres

50 miles

100 km

Fig 67

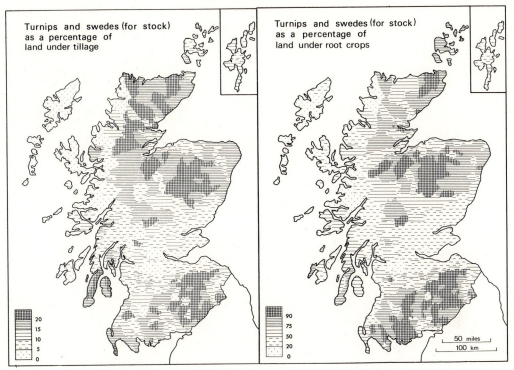

Figs 68–69

Turnips and Swedes for Stockfeeding

Area: 186,557 acres (75,500ha) in 1965; 136,566 acres (55,268ha) in 1972

Percentage of tillage under turnips: Mean 12·9; SD 6·5; Max 37·6; Min 0·0

Percentage of roots under turnips: Mean 60·4; SD 26·8; Max 99·0; Min 0·0

Turnips and swedes have long been important fodder crops in Scotland, and a larger acreage is grown than in England despite the fact that the acreage under tillage is seven times as large in the latter country; in both England and Scotland, however, the acreage has been falling steadily. The crops are largely consumed on the farms on which they are grown, for the most part being fed off to sheep while they are still in the ground. Turnips thrive under cool moist conditions and have long been part of the traditional rotation of oats, turnips, oats, followed by leys (Fig 67).

They are widely grown throughout eastern and southern Scotland and are particularly prominent in the north-east, where they are grown on most farms which have some tillage. In relative terms, they are particularly a crop of the upland margins, accounting in 1965 for a fifth or more of the tillage acreage on the higher ground of the Tweed basin and of north-east Scotland (Fig 68); the large blocks in north Scotland are more a reflection of parish shapes than of the importance of the crops (cf Fig 1). The predominance of turnips and swedes over other root crops in these localities was even more marked; over most of the Southern Uplands and the hill country of the north-east the percentage was 90 or more, and over most of the remaining uplands 75 or more (Fig 69). The main exception was along the west coast and on the islands (except Orkney), where the acreages under all crops were small and potatoes the preferred root crop.

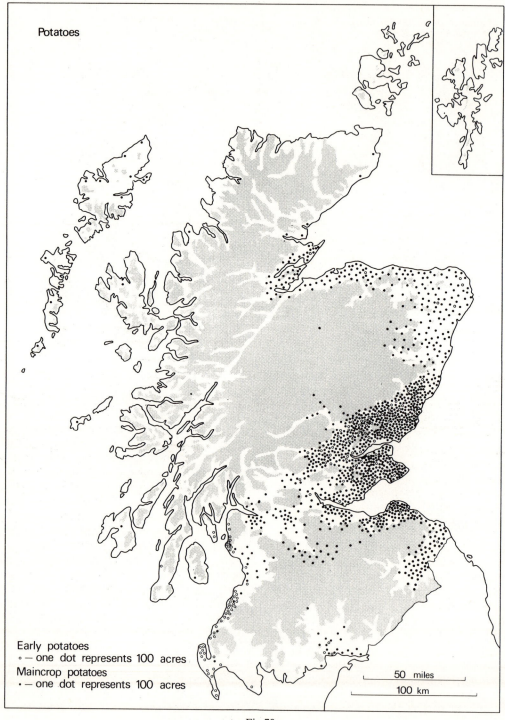

Potatoes

Early potatoes
∘ — one dot represents 100 acres
Maincrop potatoes
• — one dot represents 100 acres

50 miles

100 km

Fig 70

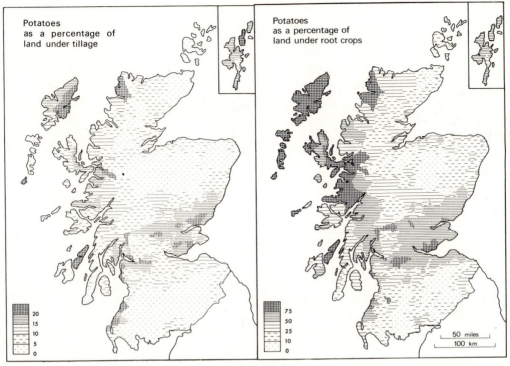

Figs 71-72

POTATOES

Area: 141,668 acres (57,333ha) in 1965; 90,452 acres (36,606ha) in 1972

Percentage of tillage under potatoes: Mean 8·3; SD 6·8; Max 39·2; Min 0·0

Potatoes were the leading cash crop in Scotland in 1965 accounting for 6·7 per cent of agricultural output and 36·0 per cent of that from crops, but their place has since been taken by barley. Potatoes are grown both on a small scale for domestic use and increasingly by large growers as a sale crop. Two-thirds of those occupiers with any land under crops in 1965 grew potatoes, but only about half of these were commercial producers; of the latter, 520 producers with 50 or more acres each had more than a third of the acreage, while those growing between 20 and 50 acres had a further third. Such concentrations are becoming more marked and the number of registered producers is declining. Potatoes are unusual among crops in that the acreage planted is controlled by the Potato Marketing Board, with which growers of one acre or more of potatoes for sale must register; registered growers have a basic acreage, which they may let to other growers, and the Board may prescribe what proportion of basic acreages may be planted in any year.

Potatoes are widely grown, but most of the acreage grown in 1965 was concentrated in the eastern half of the central lowlands, particularly in Angus and Perth, the leading counties, with 18 and 15 per cent respectively, and in Fife and East Lothian (Fig 70). Perth and Angus were also the principal areas when the relative importance of potatoes is considered, with 15 per cent or more of the tillage under potatoes in 1965, but the west coast, including both those areas in south-west Scotland specialising in early potatoes and the largely subsistence production of the crofting areas, was also important (Fig 71). In the major growing areas, too, potatoes comprised at least half the root crops, but the highest values were to be found in the north-west mainland and on the islands, where 75 per cent or more of the small acreages of roots were under potatoes (Fig 72).

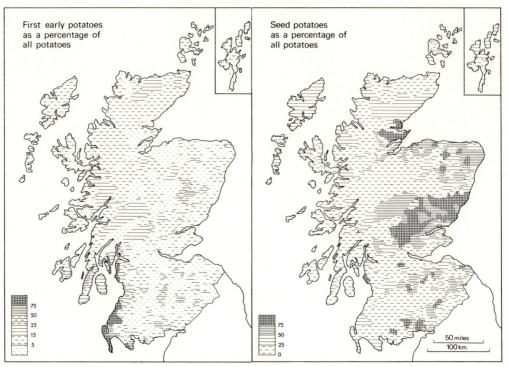

Figs 73–74

Potatoes comprise three crops, although the dividing lines between them are not sharp: early potatoes, normally harvested by the end of July; maincrop potatoes for human consumption, mainly harvested in October; and potatoes of both kinds grown for seed, either for use in Scotland or for export to England and Wales. Of the total tonnage in the 1960s, about 20 per cent was sold for seed in England and Wales, 15 per cent was used for seed on Scottish farms and a further 5 per cent found other outlets for seed. About 47 per cent of the tonnage was used for human consumption (about a tenth of this being early potatoes), and the remaining 13 per cent was used for stockfeed, destroyed or disposed of in other ways. The different crops not only have different markets, but also have rather different requirements and locations.

Maincrop potatoes and second earlies accounted for 124,626 acres (50,435ha) in 1965, or 88 per cent of the acreage planted. For this reason their distribution is very similar to that of all potatoes, with the main centres of production in the eastern half of the central lowlands, particularly in the counties of Angus, Fife and Perth, with 19, 14 and 15 per cent of the acreage respectively; areas further north are more likely to have planting delayed by late frost and so incur greater risks of damage by blight and difficulties with late harvests. Crops grown for human consumption account for about half the acreage of maincrop potatoes, the other half being grown for seed. Although the main areas of production are the same, a higher proportion of ware potatoes is grown in south-east and south-west Scotland, which together account for about a third of the acreage; for proximity to markets is of some advantage for so bulky a crop as potatoes. Maincrop potatoes grown for seed, by contrast, are more prominent in Perth and Angus, which together account for nearly half the acreage.

First early potatoes occupied 17,042 acres (6,897ha) in 1965 and were grown mainly in the south-west and the principal potato area in east central Scotland (although some earlies are grown in all counties). This is a speculative crop which does not benefit from the price support measures of the Potato Marketing Board and is highly vulnerable to the vagaries of the weather; for prices fall rapidly as the season advances. In the south-west, the crop is intended mainly for human consumption and most is grown along a narrow coastal strip of sandy soils in Ayrshire where the growing season begins relatively early (cf Fig 13). The early crop in the principal areas for potatoes, on the other hand, is grown mainly for seed.

When first earlies (both seed and ware) are considered as a percentage of all potatoes, the predominance of the south-west, particularly the Ayrshire coast, was even more marked, with 75 per cent of the acreage of potatoes in 1965 devoted to earlies, the highest value being 99·5 per cent (Fig 73). Proportions were high elsewhere along the coasts of south-west Scotland, but in other parts of the country earlies accounted for less than 25 per cent of potatoes, and often much less.

The acreage devoted to seed potatoes has already been considered briefly in the discussion of maincrop and early potatoes, although it should be noted that the larger potatoes grown for seed may be marketed as ware. Seed potatoes are inspected by staff of the Department of Agriculture and Fisheries, which publishes a register of certified crops and prescribes conditions; in 1965 there were some 72,800 acres (30,000ha) of certified seed potatoes. This is the most distinctive aspect of potato growing in Scotland, for while about a quarter of the potato acreage in Great Britain is grown in Scotland, about two-fifths of the certified seed potatoes are grown there. About half the acreage of potatoes grown in England is planted with Scotch seed and a large part of the remainder consists of once- or twice-planted Scotch seed. The cool climate of Scotland inhibits the migration of vectors of virus diseases, and the higher grades of certified potatoes require the cool moist climate of the upland margins. Although the acreages of seed and ware potatoes are broadly similar, seed production is in fewer hands and there is an increasing tendency for merchant growers to produce a substantial part of the crop, especially of the less common varieties.

Although the principal areas for seed potatoes are those for potatoes as a whole, there is a difference in that the centre of gravity of seed production lies further north than that of ware potatoes. When seed potatoes, whether for maincrop or earlies, are expressed as percentages of all potatoes, the pre-eminence of the belt from Perth to Stonehaven is clearly defined; in 1965 more than three-quarters of the acreage grown there was devoted to seed production (Fig 74); a smaller nucleus can also be identified around the Moray Firth.

Although the demand for potatoes has stabilised, higher yields imply a diminishing acreage, and stability is not helped by the seed grower's dependence on the English grower and by the fact that Scottish and English consumers prefer different varieties.

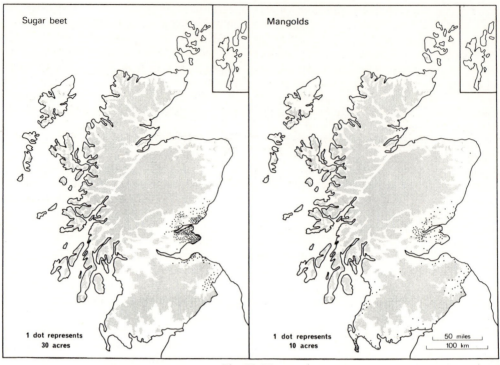

Figs 75–76

SUGAR BEET

Area: 8,647 acres (3,499ha) in 1965
Percentage of tillage under sugar beet: Mean 0·3;
SD 0·9; Max 8·0; Min 0·0

The maps of sugar beet are now only of historical interest, for a failure to grow the minimum acreage required (despite the reversal of a downward trend) led to the closure of the British Sugar Corporation's factory at Cupar (Fife), thus depriving growers of a market for their crop, most of which was grown within convenient access to the factory; in 1965, some 78 per cent of the acreage was grown in the counties of Angus, Fife and Perth (Fig 75). Nevertheless, some came from further afield, for beet was grown on contract for the British Sugar Corporation, which subsidised the cost of transport of clean beet to the factory up to a maximum of 80 miles (128km); some beet was therefore grown as far away as Berwickshire (458 acres in 1965) and Morayshire (47 acres), though northern Scotland is generally too cool for the crop. Figs 77 and 78, showing sugar beet as a proportion of the land under tillage and under roots in 1965, record the same pattern of the most highly localised farm crop grown in Scotland.

MANGOLDS

Area: 2,691 acres (1,089ha) in 1965
Percentage of tillage under mangolds: Mean 0·1;
SA 0·3; Max 1·8; Min 0·0

The maps of mangolds are similarly of historical interest only, since the crop is no longer separately recorded, but included with other fodder crops. The mangold is a succulent crop related to sugar beet and is used primarily as winter feed for cattle; it is better suited to warmer and sunnier climates than prevail in Scotland. In 1965, its distribution was similar to that of sugar beet, though some was also grown in coastal areas in the south-west (Fig 76). Figs 79 and 80, recording mangolds in relation to tillage and other root crops respectively, reveal a distribution around the coasts of south and central Scotland.

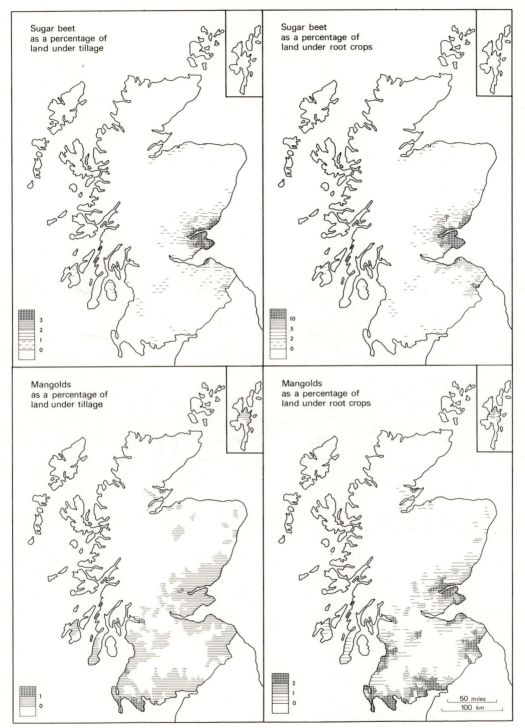

Figs 77–80

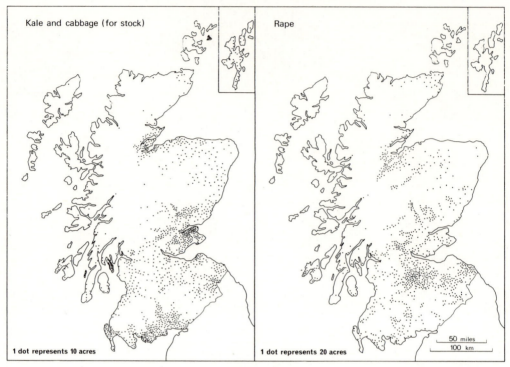

Figs 81–82

Kale and Cabbage for Stockfeeding

Area: 11,928 acres (4,827ha) in 1965; 8,726 acres (3,531ha) in 1972

Percentage of tillage under kale: Mean 1·4; SD 1·9; Max 12·5; Min 0·0

Kale and cabbage for stockfeeding were grown throughout the lowlands in 1965 (Fig 81), but were relatively most important in the comparatively mild climate of the south-west (Fig 83).

Rape

Area: 27,440 acres (11,105ha) in 1965; 24,516 acres (9,922ha) in 1972

Percentage of tillage under rape: Mean 5·1; SD 8·5; Max 100; Min 0·0

Rape, mainly for sheep feed, was also widely grown and is primarily a crop of the upland margins (Fig 82); it is frequently grown as a catch crop, or as a pioneer crop in land reclamation, and is tolerant of a wide range of conditions. In 1965 it was relatively most important in the uplands, though the acreage grown there was generally small (Fig 84).

Mashlum, Beans, Rye, Vetches and Other Green Fodder Crops

Area: 9,087 acres (3,678ha) in 1965; 7,607 acres (3,079ha) in 1972

Percentage of tillage under mashlum etc: Mean 1·1; SD 2·9; Max 35·6; Min 0·0

These diverse crops were widely grown in small acreages, but were most prominent in the livestock rearing and feeding areas of the north-east; in relative terms, however, they were most important in the uplands and along the west coast (Fig 85).

Bare Fallow

Area: 13,615 acres (5,510ha) in 1965; 9,096 acres (3,681ha) in 1972

Percentage of tillage under bare fallow: Mean 2·2; SD 6·3; Max 63·3; Min 0·0

Bare fallow is the most variable component of the area under tillage, and with the advent of herbicides, is much less extensive than formerly. In 1965 it was both absolutely and relatively most important in the Highlands (Fig 86), but this was not so in 1972.

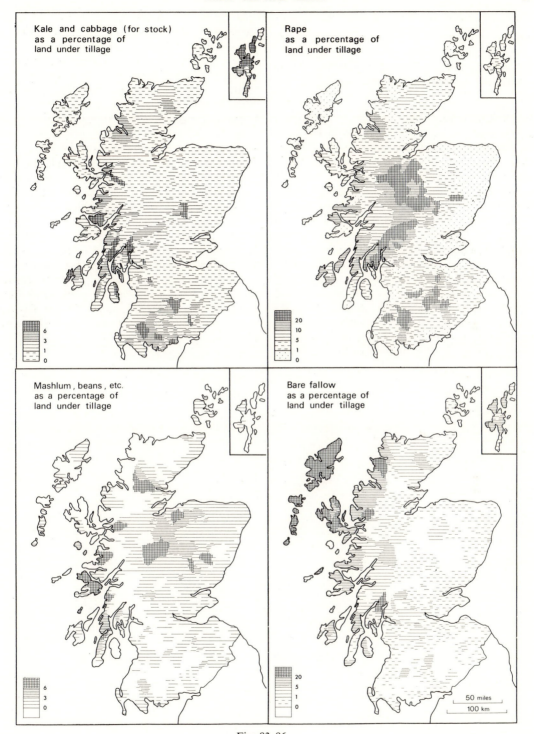

Kale and cabbage (for stock)
as a percentage of
land under tillage

6
3
1
0

Rape
as a percentage of
land under tillage

20
10
5
1
0

Mashlum, beans, etc.
as a percentage of
land under tillage

6
3
1
0

Bare fallow
as a percentage of
land under tillage

20
5
1
0

50 miles
100 km

Figs 83–86

Vegetables for Human Consumption

Area: 11,646 acres (4,713ha) in 1965; 14,486
    acres (5,862ha) in 1972
Percentage of tillage under vegetables: Mean
    0·54; SD 1·72; Max 17·5; Min 0·0

Horticultural crops accounted for 3·8 per cent of
agricultural production in Scotland in 1965 and
4·7 per cent in 1970; vegetables accounted for
about half of this. Consumption of fresh
vegetables per head of population is lower in
Scotland than in England, and the acreage under
vegetables is only some 3 per cent of that in
Great Britain, though Scotland has 13 per cent
of the land under tillage. The reasons are
complex, for considerable quantities of vegetables
are sent to central Scotland from England,
especially from the Fenland, while there are said
to be large areas of land suitable for vegetable
growing in Scotland; temperature is a major
constraint and the range of vegetables grown is
much smaller. Partly because of improvements
in transport and partly because of the increasing

proportion of horticultural crops which is
processed, the importance of proximity to
urban markets has been steadily declining.

There were two principal areas for vegetable
growing in 1965, namely, Midlothian and East
Lothian and the lowlands between Perth and
Stonehaven; much smaller acreages were grown
around Glasgow and the Moray Firth lowlands
(Fig 87). Vegetables are also grown by a very
small minority of occupiers, only 4 per cent of
those with any crops growing vegetables in 1967.
Vegetable growing to the east of Edinburgh is
more intensive, with an emphasis on crops for
consumption as fresh vegetables; in Angus,
Kincardine and Perth, the proportion of the
tillage acreage under vegetables was generally
less than 5 per cent and most of the vegetables
are grown by, or on contract for, processors,
generally on a large scale (Fig 88). Moreover,
whereas vegetable growing is long established
in the Lothians, the area north of the Tay has
risen to prominence only since the 1920s, with
the coming of canning factories and, in the
1940s, freezing plant.

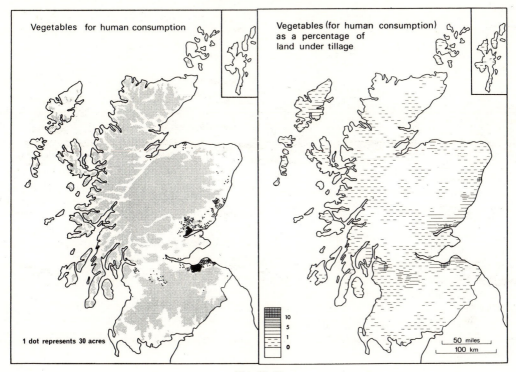

Vegetables for human consumption

1 dot represents 30 acres

Vegetables (for human consumption)
as a percentage of
land under tillage

10
5
1
0

50 miles
100 km

Figs 87–88

Peas for Canning, Quick-freezing and Harvesting Dry

Area: 4,141 acres (1,676ha) in 1965; 4,477 acres (1,812ha) in 1972

Peas for processing are the most important vegetable by acreage, accounting for nearly a third of the total; yet production is highly mechanised and the crop was grown by less than a hundred growers, with 68 growers with 30 acres (12ha) or more in 1969, accounting for 89 per cent of the acreage. Peas were grown mainly around Dundee and Montrose, with the counties of Angus, Kincardine and Perth accounting for over 99 per cent in 1965 (Fig 89); this pre-eminence has since declined, with Fife accounting for 27 per cent in 1972 and Berwick and Roxburgh for 21 per cent. Little is grown for harvesting dry, and the location of factories is an important determinant, especially for peas for quick-freezing; the expansion in Fife is largely a consequence of the closure of the Cupar sugar beet factory. The acreage of green peas for market is very small.

Broad Beans for Canning, Freezing or Drying

Area: 169 acres (68ha) in 1965; not separately recorded in 1972

This crop is also grown for processing, and in 1965 nearly all the acreage was grown near Montrose (Fig 90).

Cauliflower and Broccoli

Area: 762 acres (308ha) in 1965; 892 acres (361ha) in 1970

By contrast, cauliflower and broccoli, grown almost entirely for consumption as fresh vegetables, were highly localised in Midlothian and East Lothian; a large proportion of the acreage was in the hands of a few growers (Fig 91). There was a subsidiary area in the Clyde Valley. These crops are frost sensitive and tend to be grown in coastal locations.

Turnips and Swedes for Human Consumption

Area: 700 acres (283ha) in 1965; 1,040 acres (421ha) in 1972

The Lothians, with almost half the acreage, were the principal areas in 1965 for these crops also, which are more widely grown than cauliflower, with less important areas in Strathmore, the Clyde valley and around Aberdeen (Fig 92).

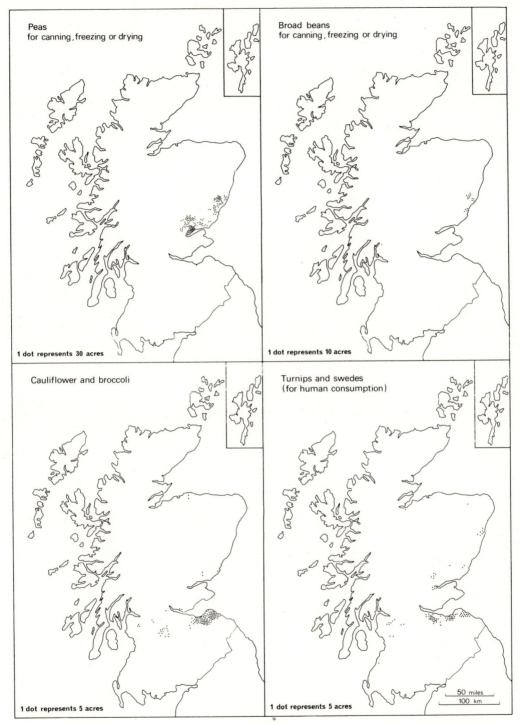

Figs 89–92

BRUSSELS SPROUTS

Area: 911 acres (369ha) in 1965; 1,324 acres (536ha) in 1972

Brussels sprouts have a long season and are grown mainly for sale as fresh vegetables. The crop was mainly grown in East Lothian and Midlothian in 1965, with these two counties providing two-thirds of the crop (Fig 93); the lowlands north of the Tay were a secondary area. The acreage has since risen markedly, and the degree of concentration has increased. Much of the crop is grown on a large scale, with ten growers in 1969 having 62 per cent of the acreage.

CABBAGES AND SAVOYS

Area: 1,618 acres (655ha) in 1965; 1,937 acres (784ha) in 1972

Cabbages and savoys were the second crop by acreage after peas and were also grown mainly in Midlothian and East Lothian, which had over 70 per cent of the crop in 1965, though smaller acreages were scattered throughout the lowlands (Fig 94). These vegetables have been losing ground to more valuable crops since World War II and the acreage under cabbages has almost halved since 1939.

CARROTS

Area: 823 acres (333ha) in 1965; 1,948 acres (788ha) in 1972

The acreage under carrots has been rising rapidly and had overtaken that of cabbages in 1972; most of the crop is now processed. In 1965 it was grown mainly east of Dundee and in Strathmore, with Angus accounting for 39 per cent of the crop, and on the light sandy soils of the coast of Moray; smaller acreages were grown in East Lothian and Fife (Fig 95). With the increasing importance of processing, Angus's share of the acreage had risen to 52 per cent in 1972, but the dominance of large growers is less marked than with peas.

LETTUCE

Area: 521 acres (211ha) in 1965; 483 acres (195ha) in 1972

Lettuce accounts for about 3 per cent of the vegetable acreage and is grown primarily around Edinburgh and Glasgow, mainly by small growers (though eleven produced a third of the crop in 1969); it is grown entirely for the fresh market. In 1965 and 1972 a third of the crop was grown in Midlothian and East Lothian and a further third in Lanarkshire (Fig 95).

OTHER VEGETABLES

Apart from tomatoes grown under glass, the other principal vegetables were beetroot, grown mainly for processing, with 59 per cent of the 244 acres (99ha) in 1965 and 58 per cent of the 383 acres (155ha) in 1972 being grown in the Lothians; leeks, recorded in the winter census, and also grown mainly in the Lothians, which had 75 per cent of the acreage of 268 acres (108ha) in 1965 and of 382 (155ha) in 1972; and rhubarb, with two-thirds of the 574 acres (232ha) in 1968 and of the 501 acres (203ha) in 1972 being grown in Lanarkshire and Renfrewshire, and with ten growers accounting for three-quarters of the crop in 1969.

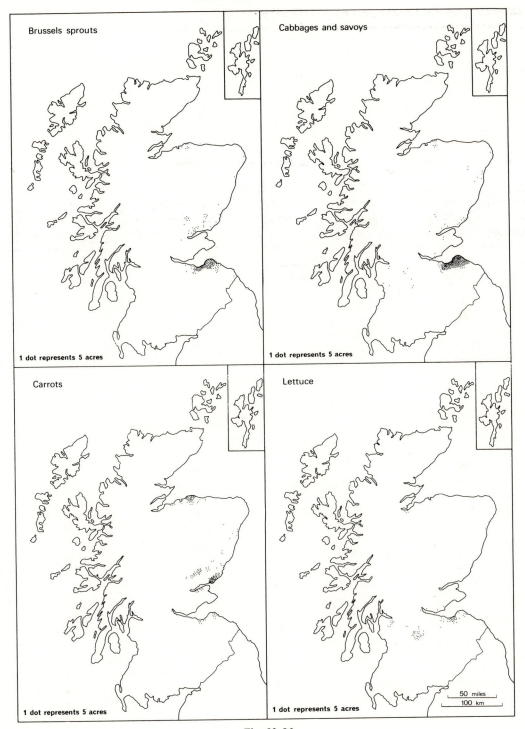

Figs 93–96

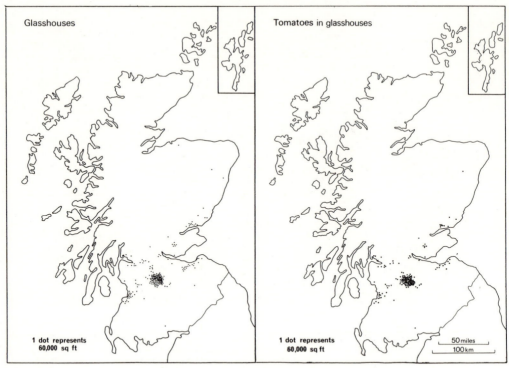

Figs 97–98

GLASSHOUSES

Area: 12·9 million sq ft or 296 acres (120ha) in 1965; 271 acres (110ha) in 1972

Glasshouses account for about 30 per cent of Scottish horticultural output; more than one crop is taken from part of the acreage, so that the cropped acreage is about 1½ times that recorded in June. Lanark accounted for about half the acreage under glasshouses, though for only a quarter of holdings with glass, and smaller acreages are scattered throughout the central lowlands, mainly in close proximity to towns (Fig 97). The glasshouse industry originated in the nineteenth century when cheap coal and close proximity to the growing Clydeside market were important; the glasshouses recorded in Lanark are highly localised in the Clyde valley south of Hamilton. While the area under glass has tended to decline, that which remains is far better equipped and is being replaced at the rate of about 10 acres a year.

TOMATOES

Area: 9·3 million sq ft or 214 acres (87ha) in 1965; 188 acres (76ha) in 1965

Tomatoes were the principal vegetable grown under glass and the distributions of tomatoes and glasshouses were virtually identical; tomatoes accounted for about half the cropped acreage and about three-quarters of the acreage in June. Lanark was the principal county, with over half the acreage, mainly in the Clyde Valley (Fig 98); less important areas included the coast of Ayr and around Dundee and Edinburgh. Disadvantages of higher heating costs and lower sunshine compared with southern England seem to be offset by proximity to markets and local reputation.

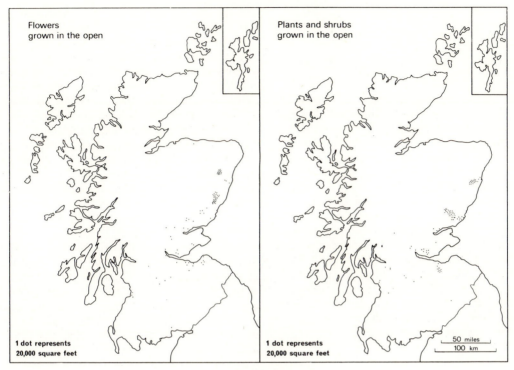

Figs 99–100

FLOWERS GROWN IN THE OPEN

Area in 1965: 2·25 million sq yd or 465 acres (188ha)

Flowers and nursery stock together account for about a quarter of Scottish horticultural output. The leading counties for flowers grown in the open in 1965 were Aberdeen and Kincardine, with over two-fifths of the acreage (Fig 99), but there were lesser concentrations elsewhere, notably around Edinburgh; many of the small producers of cut flowers also have glasshouses or grow vegetables intensively. In addition to the acreage of non-bulb flowers cut in the open, a much smaller acreage of flowers cut under glass was grown in the same areas, and 168 acres (68ha) of bulbs were grown for sale as bulbs by only a small number of specialist producers, mainly in Aberdeen, Angus and Kincardine, which had 87 per cent of the acreage.

PLANTS AND SHRUBS GROWN IN THE OPEN

Area in 1965: 2·26 million sq yd or 467 acres (189ha)

Some 10 per cent of horticultural output comes from the sale of hardy nursery stock, much of the acreage being in the hands of large growers, with the ten largest accounting for over half the crop in 1970. The distribution of land under plants and shrubs for sale in 1965 was very similar to that of cut flowers, with over half the acreage being grown in Aberdeen, Angus and Kincardine (Fig 100); other centres were near Edinburgh and Ayr. The acreage has been expanding fairly rapidly in recent years.

Soft Fruit

Area: 9,616 acres (3,892ha) in 1965; 11,359 acres (4,597ha) in 1972

Percentage of tillage under soft fruit: Mean 0·38; SD 1·37; Max 11·3; Min 0·0

Soft fruit, particularly raspberries, are one of the few horticultural enterprises in which Scotland has a more than proportionate share of British production. Soft fruit contribute about a quarter of Scottish horticultural output and, whereas the acreage in England and Wales has been declining in the post-war period, that in Scotland has remained fairly stable and now accounts for about a quarter of the acreage in Great Britain. The principal areas in 1965 were in Perth and Angus, which had 79 per cent of the Scottish acreage, particularly around Blairgowrie, Dundee and Forfar (Fig 101); there were much smaller concentrations east of Edinburgh, in the Clyde valley and around the Moray Firth, though some soft fruit was grown in many parishes (Fig 102).

Raspberries

Area: 6,908 acres (2,796ha) in 1965; 8,323 acres (3,368ha) in 1972

Percentage of soft fruit under raspberries: Mean 12·3; SD 26·8; Max 100·0; Min 0·0

This prominence of soft fruit growing in Scotland is largely due to the raspberry crop, for some five-sixths of the acreage under this crop in Great Britain is grown in Scotland, and raspberries in turn account for nearly three-quarters of the acreage under soft fruit. Their distribution is therefore very similar to that of soft fruit generally, except that it is more localised in Perth and Angus, which had 92 per cent of the acreage in 1965 and 85 per cent in 1972 (Fig 103). The underlying reasons for this concentration are primarily climatic, for raspberries grow best where summers are rather cool and there are no hot drying periods; but reputation and the inertia provided by processing plant (for most of the crop is made into jam, and some is canned or frozen) help to maintain this degree of concentration. Much of the crop is grown by relatively large producers and in 1965 over half the

acreage was grown by fifty-seven growers with 25 or more acres (10ha) under raspberries. Outside the main growing areas there are numerous growers, but they account for a very small part of the acreage. Furthermore, although much of the strawberry crop is also grown where raspberries are important, the latter accounted for three-quarters or more of the acreage under soft fruit in all the main areas (Fig 104).

The rise of Perth and Angus is primarily due to the existence of jam factories and to the development of a canning (and later quick-freezing) industry; indeed, the development of other horticultural crops for processing is said to have followed the creation of facilities for dealing with the raspberry crop. The individual stimulus appears to have been a local initiative in the Blairgowrie area (from which raspberry growing spread outwards), where two companies were formed to purchase farms, which were then subdivided into smallholdings. The coincidence of the harvest and the Scottish school holiday period is also said to have been important, although this would have applied to any Scottish location with reasonable access to a large population, and is rather a factor helping to explain why the crop has flourished in Scotland rather than in England.

Strawberries

Area: 1,802 acres (729ha) in 1965; 2,412 acres (976ha) in 1972

Angus and Perth are also the main areas for the strawberry crop, though their predominance is less marked and they contribute about half the acreage (though both this proportion and the acreage under strawberries have been rising). Lanark was once the principal area, with about half the acreage, but urban expansion and the ravages of disease halved the acreage in the early 1930s. Despite the smaller acreage, there are almost as many growers of strawberries as there are of raspberries, and the dominance of large growers is less marked (though the latter, often producing for processing, have a larger share of the acreage grown north of the Tay). Much more of the crop is, however, marketed fresh, for the fact that strawberries tend to ripen

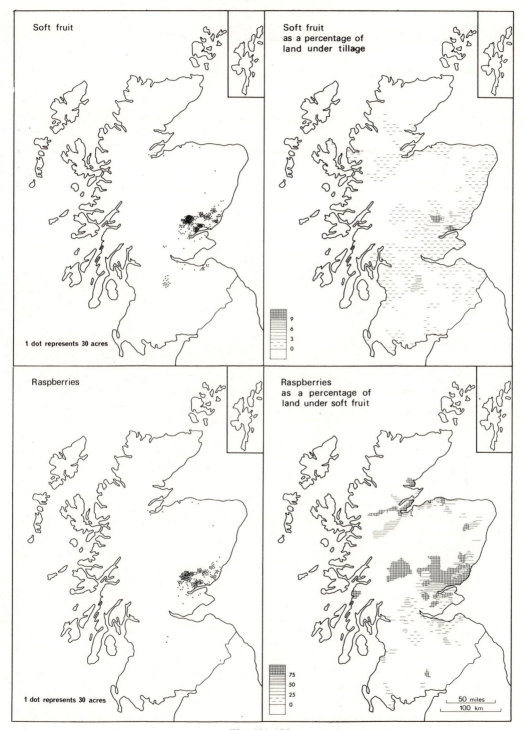

Figs 101–104

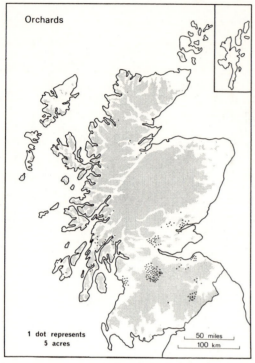

Fig 105

some three weeks later in Scotland than in southern England helps to extend the short season when fresh strawberries are available.

OTHER SOFT FRUIT

Most of the remaining acreage of soft fruit is shared between blackcurrants and gooseberries, though the acreage of both has declined markedly since before World War II; the major areas for soft fruit growing are important for both crops.

ORCHARDS

Area: 855 acres (346ha) in 1965; Area of commercial orchards in 1972: 354 acres (143ha)

Although the growing of top fruit was formerly more important, orchards have never been a major feature of Scottish farming, largely for climatic reasons. They are mainly located in the lower Clyde valley in Lanarkshire, a fact obscured in Fig 105 by the difficulty of locating a large number of dots in a small area; for the distribution is in fact linear. Other smaller concentrations were to be found in southern Perthshire, around Edinburgh and Dundee, in Teviotdale, on the Ayrshire coast and in Kintyre. Many of the orchards in these outlying areas do not, however, produce fruit for sale or manufacture, a view confirmed by the fact that while Lanark had 60 per cent of all orchards in 1968 (the date of the last comprehensive orchard census), it had 73 per cent of the acreage of commercial orchards. These accounted for 65 per cent of all orchards in 1968, though the discovery that some orchards previously returned in the census produced no fruit led to their elimination from the census to give a total acreage of 552 acres (223ha) in 1968. Production from the non-commercial orchards is generally low, for trees are often old or diseased or more widely spaced than in the commercial orchards.

Some 80 per cent of the commercial acreage was under plums and damsons in 1968 (with 56 per cent under Victoria plums alone); cooking apples were the second crop, with 15 per cent of the acreage. The predominance of plums and damsons was even more marked in Lanark (88 per cent) and East Lothian (92 per cent), but these crops accounted for only 32 per cent of the commercial acreage in Perthshire, where cooking apples were the principal crop; over 40 per cent of the commercial acreage under plums and damsons was under-planted with other crops, mainly gooseberries. Plums and damsons occupied only 44 per cent of the non-commercial acreage, and were followed closely by cooking apples, with 40 per cent. Growing is predominantly on a small scale, with only 19 out of the 122 holdings with commercial orchards in 1968 having more than 5 acres of orchards, although they accounted for 53 per cent of the acreage; at the other extreme, fifty-one growers had 1 acre or less. Half of these commercial orchards were on full-time horticultural holdings, and half the remainder were part-time holdings which were also predominantly horticultural; most other commercial orchards were on cropping farms.

Both the extent and the composition of Scotland's orchards have changed considerably over the past seventy years, though Lanark has long been the leading county. The acreage of orchards was at a maximum in 1904–6, when there were some 2,500 acres (1,000ha), and by 1939 there were some 1,330 acres (538ha), of which 45 per cent were under apples (compared with 28 per cent in 1968) and 25 per cent under plums (compared with 67 per cent); Lanark was the leading county, with 37 per cent of the acreage, followed by Perth with 29 per cent. It seems likely that the acreage will further decline and become more localised in Lanarkshire, for while only 12 per cent of trees on commercial orchards were less than ten years old in 1968, the proportion was higher in Lanarkshire and reaching 20 per cent of the acreage under Victoria plums.

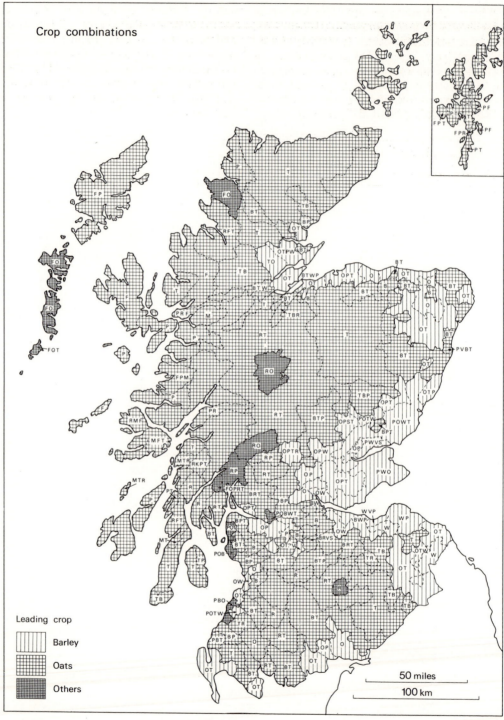

Crop combinations

Leading crop

	Barley
	Oats
	Others

50 miles
100 km

Fig 106

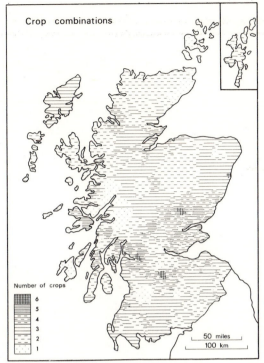

Crop combinations

Number of crops

6
5
4
3
2
1

50 miles
100 km

Fig 107

different crops, ranked in descending order, and identifying (by the method of least squares) those combinations of crops which most closely resemble ideal combinations in which the component crops occupy equal shares of the acreage under tillage, eg 50 per cent each for two crops, $33\frac{1}{3}$ per cent for three crops and 25 per cent for four crops; when the combination has been identified, all other crops are then ignored. The crops which appear in Fig 104 are those identified in this way; the highest ranking crop in each combination is shown by shading and other crops in descending rank order, by letters. The crops identified are: B, barley; F, fallow; K, kale and cabbage; M, mixed grain; O, oats; P, potatoes; R, rape; S, soft fruit; T, turnips and swedes; V, vegetables; and W, wheat. Fig 106 shows the number of crops appearing in the crop combinations as identified.

As would be expected from the use of acreages as the basis of comparison, cereals were the dominant crops, barley being the leading crop throughout the eastern lowlands and in parts of the south-west, and oats nearly everywhere else. In the lowlands another cereal (generally wheat in the south and oats in the north) was usually the second crop and a cereal was often the third crop; the regional contrasts in cropping between the Merse, the Lothians, the lowlands north of the Tay and north-east Scotland are also apparent. In the uplands, second and third ranking crops, where they existed, were generally rape or turnips. No clear pattern of cropping is revealed by this map or by Fig 107, though it will be appreciated that the acreages of crops in those areas where oats are the leading crop are small; numbers of crops are generally highest in and around the central lowlands.

CROP COMBINATIONS

The individual crops described in the preceding pages are not grown in isolation, but rather in various permutations with other crops. Rotations are now much more flexible than formerly, but it is none the less useful to consider cropping as a whole. Fig 106 represents an attempt to do so. It has been constructed by comparing the proportions of the acreage of tillage under

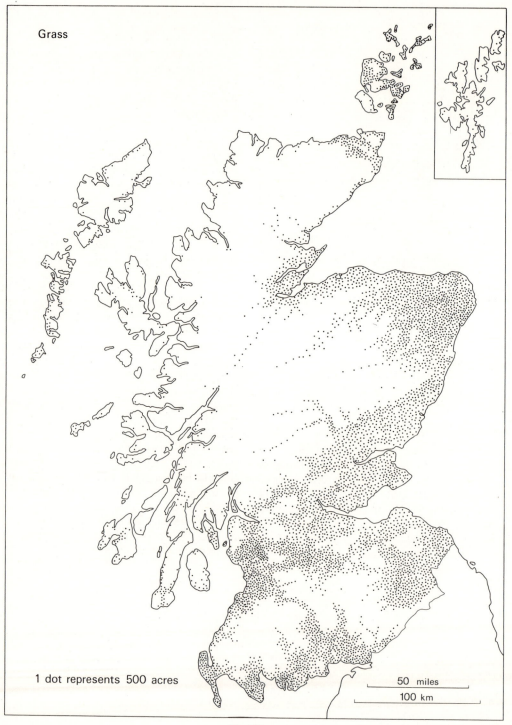

Grass

1 dot represents 500 acres

50 miles

100 km

Fig 108

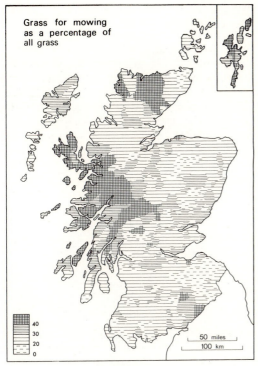

Grass for mowing
as a percentage of
all grass

40
30
20
0

50 miles
100 km

Fig 109

Grass

Area: 2,785,162 acres (1,127,155ha) in 1965;
2,714,076 acres (1,098,387ha) in 1972

In terms of area, grass is the most important
crop grown on Scottish farms, accounting for
65 per cent of the area under crops and grass in
both 1965 and 1972 and providing rather more
than half of the total supplies of animal feeding
stuffs in 1967/8. Yet surprisingly little is known
about it. Unlike England and Wales, there was
no survey of Scottish grassland in the 1940s; and
although the Agricultural Survey of Scotland in
1941 did provide some information, this was
primarily concerned with the potential for up-
grading permanent grass to temporary grass and
rough grazing to permanent or temporary grass.
Although a distinction was formerly made in the
agricultural census between permanent grass,
which is not part of the arable land and is rarely
if ever ploughed, and temporary grass, which
forms part of rotations, there were many
problems and differences of interpretation, and

the distinction was abandoned for census purposes in 1959. For comparisons with other parts
of the United Kingdom, however, grass under
seven-years old and that seven-years old and
over are equated with temporary and permanent
grass respectively.

Fig 108 shows the distribution of all grassland
in 1965, ie the grassland component of crops
and grass and excluding rough grazing. Grass
was widespread throughout the lowlands, though
there was clearly more grass in the lowlands of
the south-west. The proportion of improved land
in grass is, of course, the mirror image of Fig 52,
which records the proportion of improved land
in tillage. The lowest values, under 50 per cent,
were in the Merse, the Lothians, east Fife and
Strathmore; in the north-east, from Aberdeen
to Caithness, grass generally occupied between
50 and 75 per cent; and in the west and south-
west the proportion was higher, though in few
areas, even in the uplands, was it greater than
90 per cent. It must, of course, be appreciated
that such grass may differ greatly in quality;
trials on grassland have shown a seven-fold
difference in carrying capacity between fattening
pastures and poor, agrostis-dominated permanent pasture, and it was estimated in 1941
that nearly 1 million acres (400,000ha) of
permanent grass were unsuitable for cropping.

Grass for Mowing

Area: 726,809 acres (294,140ha) in 1965; 807,146
acres (326,652ha) in 1972
Percentage of grass for mowing: Mean 28·0; SD
8·3; Max 68·0; Min 6·4

On average, 24 per cent of the grass was mown
in 1965 and 30 per cent in 1972, and just over a
fifth of the contribution of grass to supplies of
animal feeding-stuffs was estimated to be due to
hay and silage in 1967/8. While most of the grass
acreage was in the lowlands, especially in the
south-west, the highest percentages of mown
grass were in the western Highlands and the
islands, notably the inner Hebrides and Shetland
(though the acreages grown were in fact small).
These are the areas with the highest rainfall and
the least favourable conditions for making hay;
but they are also the areas with least tillage,
where preserved grass is an important component

of winter feed. The range of percentages is not, however, very great, being between 20 and 30 per cent in most of the eastern lowlands where there are other sources of winter feed and much of the grassland is required for fattening sheep and cattle (Fig 109)

MOWN GRASS

Most grass is required for grazing rather than mowing, and even the fields which are mown provide grazing before they are closed to stock and after the grass has been cut. Yet mown grass provides four products, hay, silage, seed and dried grass, and Figs 110 to 113 illustrate their rather different distributions in 1965. This information is collected at the December census and is based on the preceding hay harvest.

HAY

Area: 540,000 acres (218,566ha) in 1965; 520,311 acres (210,570ha) in 1972
Tonnage made: 1,056,000 tons in 1965; 1,155,988 tons in 1972

The greater part of the grass cut is used for hay; in 1965 the proportion was 70 per cent and in 1972, 64 per cent, for hay making is losing ground to silage making. Grass cut for hay was widespread in 1965, particularly in the central lowlands, the south-west and the north-east (Fig 110). Over most of the uplands, the proportion of mown grass cut for hay was between 70 and 80 per cent, but in the north-east it was less than 60 per cent; in Stirlingshire, long famed for its hay crops, and in Berwickshire, the proportion exceeded 80 per cent.

SILAGE

Area: 202,089 acres (81,785ha) in 1965; 292,085 acres (118,207ha) in 1972

Tonnage made: 1,351,000 tons in 1965; 2,419,545 tons in 1972

The acreage of grass cut for silage is considerably smaller, though it is rising. Its distribution was also widespread in 1965, but north-east Scotland was clearly the most important area (Fig 111). This view is confirmed by calculation of the proportion in each county, which shows that this exceeded 40 per cent in Aberdeen and was relatively high throughout the north-east; it was also high in Argyll, though the acreage cut was small. However, as the tonnages quoted show, comparison by acreage can be somewhat misleading, as the tonnage of silage made has risen rapidly and exceeds that of hay. In addition to the grass silage illustrated here, a much smaller amount of arable silage is also made.

GRASS SEED

Area in 1965: 17,686 acres (7,158ha)

The area under grass seed in 1965 was much smaller. Its distribution is very fragmented, including such diverse areas as the Moray Firth, Buchan, the lowlands of Stirlingshire, and the coastal lowlands of Ayr and Wigtown (Fig 112). It should be noted that the acreages represented by a single dot in Figs 110 and 111 are twenty-five times as large as those represented by the dots in Figs 112 and 113.

DRIED GRASS

Area in 1965: 7,298 acres (2,954ha)

The acreage of grass cut for dried grass in 1965 was even smaller, being less than 1 per cent of the total acreage mown. Its distribution was even more fragmented than that of grass for seed, but the most important areas were the islands, especially Lewis, with the Highland Region having a third of the Scottish acreage (Fig 113).

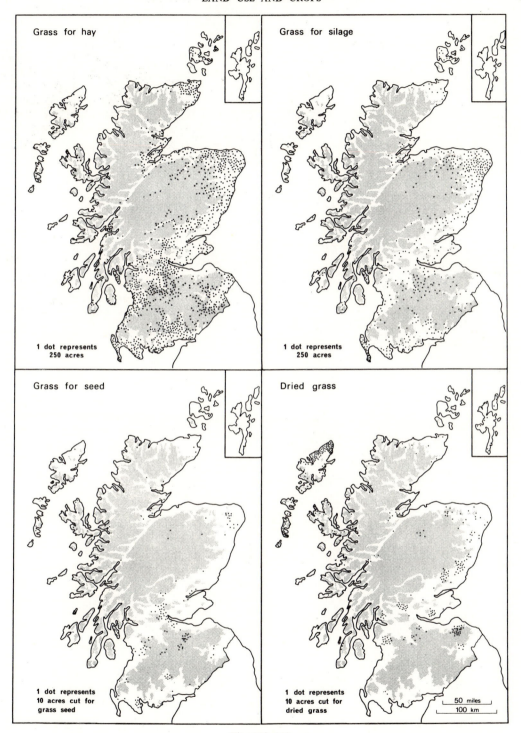

Figs 110–113

Grass (7 years and over)

1 dot represents 250 acres

50 miles
100 km

Fig 114

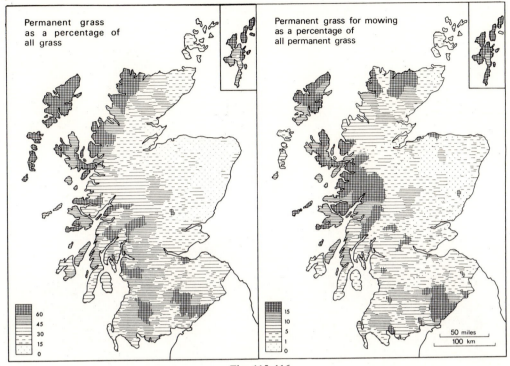

Figs 115–116

GRASS SEVEN YEARS OLD AND OVER (PERMANENT GRASS)

Area: 892,996 acres (361,396ha) in 1965; 1,049,221 acres (424,620ha) in 1972
Percentage of grass in 'permanent' grass: Mean 34·2; SD 18·8; Max 85·9; Min 1·8
Percentage of 'permanent' grass for mowing: Mean 7·4; SD 7·7; Max 51·7; Min 0·0

The area under grass which has been down for seven years or more can be broadly equated with permanent grass. In 1965, such grass was most important in southern Scotland, especially in the western half of the central lowlands and in Galloway (Fig 114). The proportion of all grass which is older grass shows a marked gradient from east to west, with the highest values in the north-west mainland and in the islands (Fig 115), though the acreages involved are small; however, proportions on the coastal lowlands of south-west Scotland were much lower than inland, and there is supporting evidence from the Land Utilisation Survey that the proportion of permanent grassland increases with elevation. In general, older grass is to be found where conditions are less suitable for cropping, on the margins of cultivation, in places where there are heavy, cold clay soils, and in areas of higher rainfall: the coincidence of the last two conditions over large tracts of western Scotland is a major factor in explaining this distribution. The Agricultural Survey of Scotland also provided evidence that the proportion of permanent grass is higher on small-holdings; on holdings with 5–50 acres of crops and grass in 1941 the proportion of that acreage in permanent grass was 56 per cent, compared with 20 per cent on holdings with 300 acres and over. Furthermore, the quality of this older grassland varies greatly with its composition, grading imperceptibly into rough grazing.

The proportion of 'permanent' grass for mowing was much smaller than that of grass as a whole, but in general it increased with the proportion of older grass (Fig 116); over most parts of the western mainland and the islands (except Orkney) the proportion was 15 per cent or over, compared with less than 5 per cent in most of eastern and north-east Scotland.

Grass (under 7 years)

1 dot represents 250 acres

50 miles

100 km

Fig 117

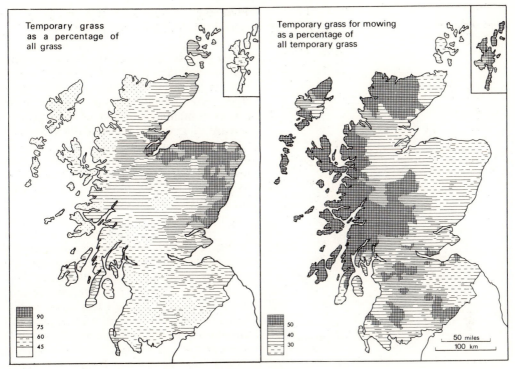

Figs 118–119

GRASS UNDER SEVEN-YEARS OLD (ROTATION OR TEMPORARY GRASS)

Area: 1,892,166 acres (765,760ha) in 1965; 1,664,855 acres (673,767ha) in 1972
Percentage of grass in 'rotation' grass: Mean 65·8; SD 18·8; Max 98·3; Min 0·0
Percentage of 'rotation' grass for mowing: Mean 40·4; SD 15·3; Max 97·0; Min 10·3

Younger grass can be broadly equated with rotation or temporary grass in other parts of the United Kingdom, though it includes a proportion of older grass which has been reseeded. It had a more easterly distribution than permanent grass in 1965, being particularly prominent in north-east Scotland, though it was an important feature of farming throughout the lowlands (Fig 117). The relative importance of temporary grass throughout the north-east, from Dundee to Inverness, but especially in Buchan, where it plays a major part in rotations, is expressed in Fig 118: over most of north-east Scotland the proportion of all grass represented by temporary grass was 90 per cent or more. Only in parts of the uplands and the islands was less than 45 per cent of the grassland temporary.

Temporary grass is generally more uniform in character and more productive than older grass and a higher proportion was intended for mowing. As on the map of permanent grass for mowing, the highest values in 1965 were to be found in the west and north-west, and in the islands, with the exception of Orkney (Fig 119); the coastal areas of the south-west had a lower proportion of mowing grass than the adjacent uplands.

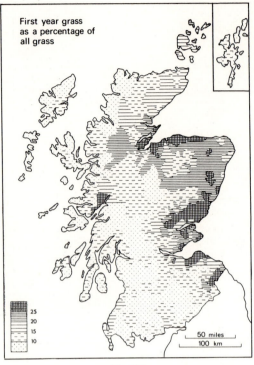

First year grass
as a percentage of
all grass

25
20
15
10

50 miles
100 km

Fig 120

FIRST YEAR GRASS
Area: 437,291 acres (176,972ha) in 1965

Percentage of grass under first year grass: Mean 15·4; SD 7·5; Max 38·2; Min 1·4

Grass leys in Scotland are sown for varying periods, and the traditional ley of the north-east was left down for three years, to be followed by three years of tillage crops. The agricultural census contains three sets of data giving the acreage of grassland sown at different times in the preceding six years. Thus, of the temporary grass recorded in 1965, 23 per cent had been sown in 1964, 40 per cent in 1962 or 1963, and 35 per cent in 1959, 1960 or 1961. Fig 120 has been included as an example of the maps that might have been constructed from this source. If longer leys were common, the acreage sown in 1962 and 1963 would be approximately twice, and that sown in 1959–61 three times the acreage sown in 1964 (with some allowance for the reseeding of older leys). Where leys are shorter, the proportion of younger leys would be higher, and Fig 120 confirms the relative (and in fact absolute) importance of younger leys in eastern Scotland. A much higher proportion of these younger leys was mown, 70 per cent, compared with 28 per cent of second and third year grass and 18 per cent of fourth, fifth and sixth year grass.

4

Livestock Farming

LIVESTOCK are by far the most important component of Scottish agriculture. Approximately 91 per cent of farmland is devoted to their support by providing grazing and preserved grass, whether as hay, silage, or dried grass, and perhaps two-thirds of the acreage of tillage crops is used to grow feeding-stuffs for livestock, the greater part of which is used on the farm of origin. Further indications of their importance are that some five-sixths of all full-time holdings in Scotland had, as their main enterprise in 1965, some form of livestock production, and that livestock and livestock products accounted for 78 per cent of agricultural production in Scotland in 1965 and 77 per cent in 1970. Indeed, these figures understate the importance of livestock in different parts of Scotland since sales of store stock, which contribute a large part of revenue on hill and upland farms in particular, are included only if such animals are sent out of Scotland. Thus, whereas some $2\frac{1}{2}$ million store lambs were sold in 1963 and almost 700,000 store calves and older cattle, the number contributing to the agricultural output of Scotland was little more than 500,000 sheep and 30,000 cattle, the remainder being sold for fattening in Scotland.

Analysis of the distribution of livestock, and more especially the assessment of their relative importance, is more difficult than that of crops, especially on the basis of census data. Livestock may be transferred from one farm to another at different stages in their lives, thus contributing to a complex network of linkages between farms in different parts of the country, of which those suggested by Figs 32 and 33 provide highly simplified and selected samples. Moreover, as Fig 12 suggests, the timing of different events in the agricultural calendar varies widely throughout the country, especially in relation to seasonal changes in the numbers of sheep of various ages. Numbers on 4 June thus give only an incomplete picture of the distribution of livestock.

Some idea of the importance of livestock in different parts of Scotland is provided by Table 13; only in the East Central and South East Regions, which contain the main cropping areas, was it under three-quarters.

TABLE 13

Livestock Output by Regions in 1961

High-land	North East	East Central	South East	South West	Scot-land
Percentage of output from livestock and livestock products					
97	78	57	65	90	75

Source: *Scottish Agricultural Economics*, Vol 15

A complementary view is given by Table 14 which records the proportion of net output in 1965 which was due to livestock and livestock products, ie sales less expenditure on seeds and feeding-stuffs, on different types of farms, together with the proportion of farms in each class (see Section 6). Only on cropping farms was the proportion less than half and only on these farms, and on arable rearing and feeding farms was it less than three-quarters.

TABLE 14

The Contribution of Livestock to Net output in 1965, by Type of Farm

Hill sheep	Upland	Rearing with arable	Arable rearing and feeding	Crop-ping	Dairy-ing
Percentage of net output					
98	94	76	61	37	87
Percentage of full-time farms					
5	13	26	9	14	29

Source: *Scottish Agricultural Economics*, Vol 17

LIVESTOCK UNITS

Number: 2,613,008 in 1965; 2,694,786 in 1972

Livestock units per 100 acres of agricultural land: Mean 24·5; SD 13·1; Max 83·6; Min 1·2

The crops and grass recorded in the agricultural census can be readily compared since they are all recorded in the same unit, viz acres; and while it can be argued that it is misleading to equate an acre of wheat and an acre of rough grazing because of the difference in productivity, such comparisons are justified, especially in terms of land use. The same cannot be said about livestock, for it is not very meaningful to compare, say, numbers of cows and those of poultry. It is therefore the usual practice to convert all the different classes of livestock into livestock units on the basis of their feed requirements. Those used in this atlas are as follows:

Dairy cows and heifers in milk and cows in calf	1
Beef cows and heifers in milk and cows in calf	$\frac{4}{5}$
Heifers in calf, bulls and bull calves, other cattle two-years old and over	$\frac{3}{4}$
Other cattle one-year old and under two	$\frac{1}{2}$
Other cattle under one-year old	$\frac{1}{4}$
Ewes and rams	$\frac{1}{5}$
Other sheep	$\frac{1}{15}$
Breeding pigs	$\frac{1}{2}$
Boars	$\frac{2}{5}$
Other pigs	$\frac{1}{7}$
Broilers and pullets	$\frac{1}{200}$
Other poultry	$\frac{1}{50}$

Numbers of horses are no longer recorded in the census, but were estimated to account for about 1 per cent of livestock units in 1967/8. While the use of such factors applied to census data inevitably implies some arbitrary decisions, the differences in the importance of the various classes of stock are sufficiently marked not to make the choice of factors of critical importance.

Fig 121 shows the distribution of livestock units in June 1965. The highest densities were to be found in the lowlands of Ayrshire and in the extreme south-west, the main dairying areas; the second most important area was Buchan. In interpreting this map some allowance must be made for seasonal differences in numbers of livestock. In particular, numbers of cattle in eastern Scotland are increased in winter by the purchase of store cattle for fattening, while store lambs are similarly transferred to the lowlands for fattening and ewe lambs for wintering. Numbers in hill areas are thus appreciably lower in winter and the contrasts between upland and lowland correspondingly greater, though there seems no reason to suppose that the broad regional contrasts are greatly modified by these seasonal differences. Numbers of livestock units are probably most stable in the main dairying areas.

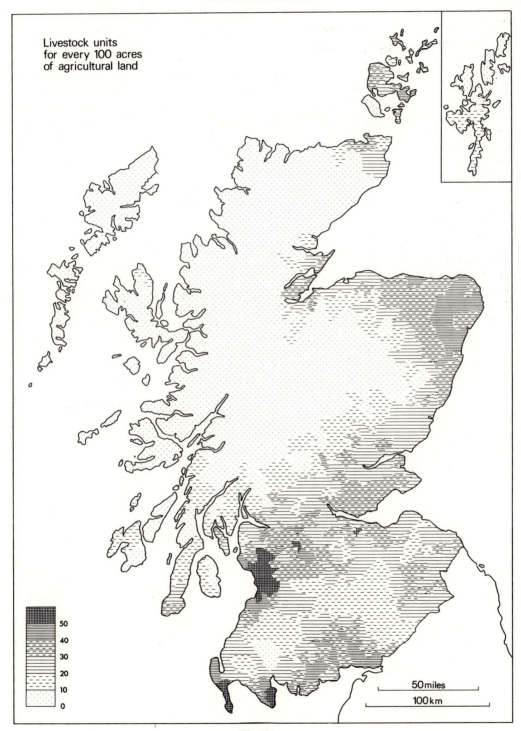

Livestock units
for every 100 acres
of agricultural land

50
40
30
20
10
0

50 miles
100 km

Fig 121

LIVESTOCK RANK

Of the 2,613,008 livestock units in 1965, beef cattle accounted for 26·6 per cent, dairy cattle for 19·7 per cent, sheep for 45·2 per cent, pigs for 4·1 per cent and poultry for 4·4 per cent; the percentages in 1972 were respectively 33·5, 17·6, 38·3, 4·5 and 6·0. While the variations throughout Scotland in the actual proportions represented by the different classes of stock are shown in the appropriate sections of this atlas, some preliminary idea of their relative importance can be obtained by comparing their rank order (Figs 122–5). The various classes of stock in each parish have been arranged in descending rank order and appropriate shadings indicate, for each class of stock in turn, whether it was first, second, third, fourth or fifth in rank in each parish. The overwhelming importance of beef cattle in relation to other livestock in north-east Scotland is clearly demonstrated in Fig 122, for beef cattle were the leading livestock in north-east Scotland and in most of the coastal lowlands and foothills from Perth to Inverness, as well as in east Fife and Orkney. Beef cattle generally ranked second in other major cropping areas and on the margins of the uplands. Dairy cattle, by contrast, were the leading livestock over all but the eastern third of the central lowlands and in Galloway and the Solway lowlands (Fig 124); they ranked third in parts of the uplands and of the Highlands and islands, and occupied the lowest rank in most of the north-east. Sheep were the leading livestock over the greater part of the Southern Uplands and the Highlands and in nearly all the islands except the Orkneys (Fig 123); they ranked second over most of the north-east and on the fringes of the uplands (though it will be appreciated from Fig 121 that densities of all classes of livestock are low in these areas). In contrast, pigs appeared as the leading livestock in only two areas (Edinburgh and Aberdeen) and occupied third or fourth rank in parts of north-east and eastern Scotland (Fig 125).

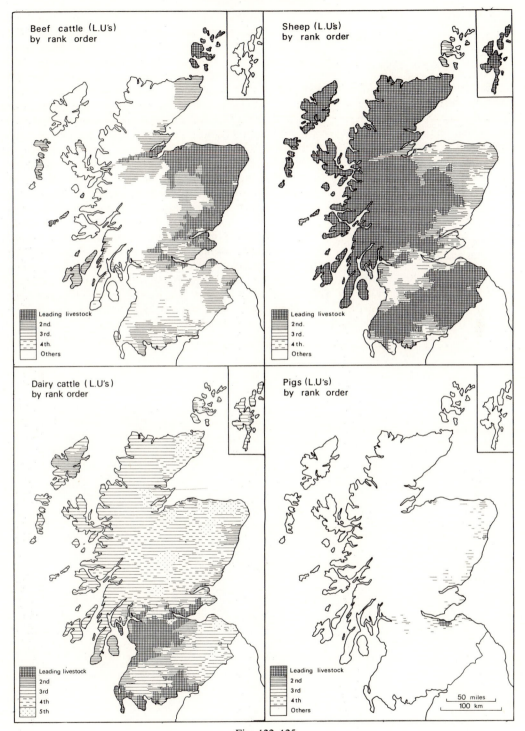

Figs 122–125

GRAZING UNITS

Number: 3,116,097 in 1965; 3,229,819 in 1972
Grazing units per 100 acres of agricultural land:
   Mean 29·5; SD 14·8; Max 68·9; Min 1·7

In one respect, viz stocking density, livestock units can be misleading, since they include not only grazing stock, but those, such as pigs and poultry which make little use of grass and generally require little land. Fig 126 has therefore been prepared to show the distribution of grazing stock, viz sheep and cattle, as expressed in grazing units. The factors used are as follows:

Cows, heifers and bulls	1
Other cattle	$\frac{3}{4}$
Ewes and rams	$\frac{1}{4}$
Other sheep	$\frac{1}{10}$

The pattern of distribution revealed by Fig 126 is very similar to that portrayed by Fig 121, a not surprising fact when cattle and sheep accounted for 91·5 per cent of all livestock units; but, although livestock units and grazing units are both calculated as densities per 100 acres of agricultural land, the two maps are not strictly comparable; not only are different factors used, but grazing accounts for only a part of the feed requirements even of grazing stock, particularly cattle. According to one estimate, the contribution of different sources of feed to the total supply of animal feeding-stuffs in 1967–8 was a minimum of 56 per cent from grazing (of which two-thirds could be attributable to grass and one third to rough grazing), 36 per cent from other home-grown feed (of which 43 per cent were cereals and concentrates, 30 per cent hay and silage and 27 per cent roots, straw etc), and 8 per cent was imported. A more useful map would be one showing the variation in stocking rates throughout the country, but this cannot be computed. It was, however, estimated in 1941 that roughly a third of all grazing stock were on rough grazing (though four-fifths of these would have been sheep).

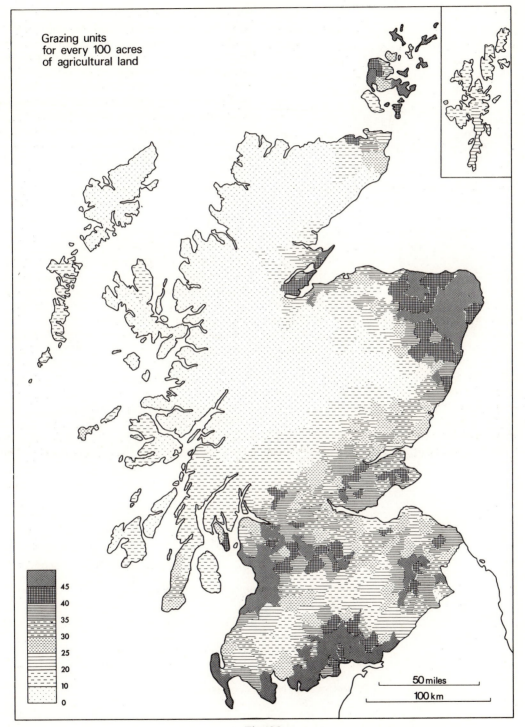

Grazing units
for every 100 acres
of agricultural land

45
40
35
30
25
20
10
0

50 miles
100 km

Fig 126

CATTLE

Number: 2,035,218 in 1965; 2,387,655 in 1972

Cattle per 100 acres of agricultural land: Mean 22·5; SD 14·8; Max 74·1; Min 0·3

Cattle per 100 acres of crops and grass: Mean 51·2; SD 20·8; Max 159·7; Min 8·0

Cattle livestock units per 100 livestock units: Mean 47·5; SD 22·1; Max 93·8; Min 1·5

Cattle grazing units per 100 grazing units: Mean 54·5; SD 25·5; Max 100·0; Min 1·6

Cattle are the most important class of livestock as measured by their contribution to agricultural output, providing through sales of animals and livestock products 58·6 per cent of output in 1965 and 58·2 per cent in 1970. Moreover, as Section 7 shows, their numbers have been increasing fairly steadily in the period since World War II. Estimates of their contribution by regions are available only for 1961 and show that cattle were important everywhere, even in the main crop growing areas, though the relative importance of sales of cattle and calves and those of milk vary considerably (Table 15), sales of milk exceeding those of cattle and calves only in South West Scotland. It is also clear that the contribution of cattle in 1961 was greatest in those regions in which cattle were most numerous.

Most Scottish farms have some cattle, the average in 1967 being 63 per cent, though this proportion would be appreciably higher if very small holdings were excluded; for the number of holdings with cattle is considerably larger than that of full-time holdings. Proportions were again highest in the two regions with the largest number of cattle and fell below 50 per cent only in the Highlands, where there are large numbers of smallholdings in the crofting districts.

Fig 127 shows the distribution of cattle in June 1965. In constructing it some discretion has been exercised in placing dots in parishes in the Highlands and Southern Uplands; they have been located on the lower slopes and in the valleys; for there are few cattle on the higher ground. Further justification for this interpretation was the finding of the Agricultural Survey of Scotland that little more than a tenth of all cattle were on rough grazings in 1941, compared with two-fifths of the sheep. The two regions already mentioned, North East and South West, stand out as the major areas, notably Buchan in the former and Ayrshire and Wigtownshire in the latter. Numbers were also high in Orkney, which again contrasted markedly with Shetland and other islands. In general, cattle were numerous throughout the lowlands and few in the uplands, a view that is independent of decisions about the location of dots.

These impressions are confirmed by Fig 128, which shows the density of cattle in 1965, calculated in relation to the acreage of all agricul-

TABLE 15

Cattle and Agricultural Output in 1961

	Highland	North East	East Central	South East	South West	Scotland
Percentage of gross output from cattle in 1961						
Cattle and calves	20	33	25	20	17	23
Milk	20	13	11	10	45	22
Total	40	46	36	30	59	45
Percentage of all cattle in each region in 1961						
	10	28	15	9	37	100
Percentage of all holdings with cattle in 1967						
	49	74	63	63	72	63

Source: Agricultural Censuses and *Scottish Agricultural Economics*, Vol 15

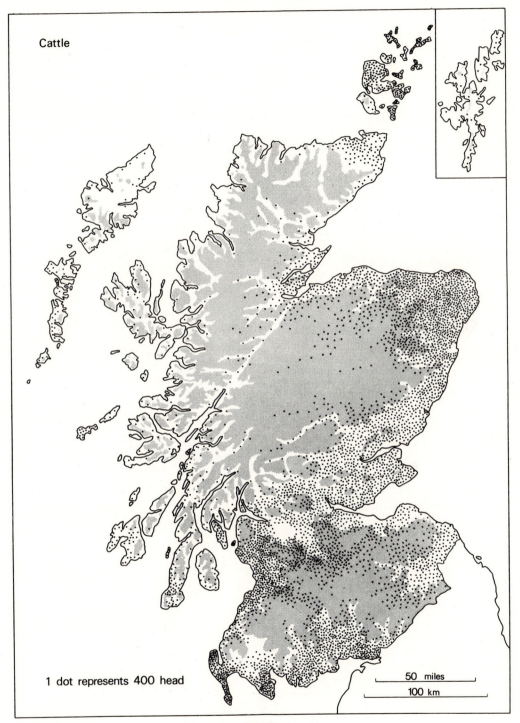

Cattle

1 dot represents 400 head

50 miles
100 km

Fig 127

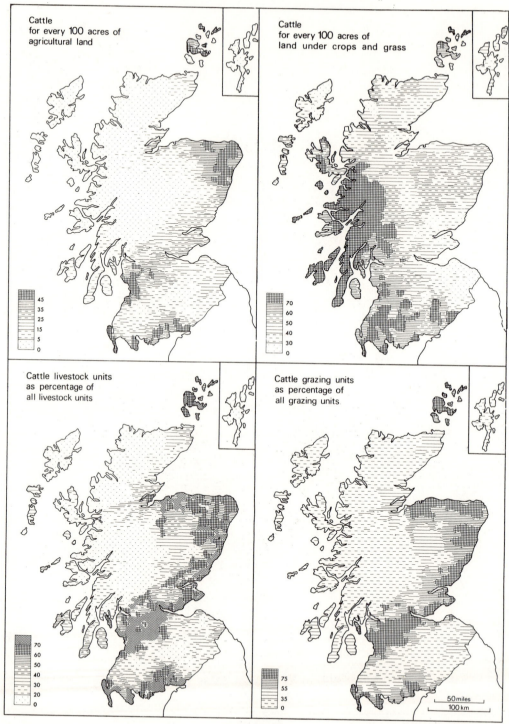

Figs 128–131

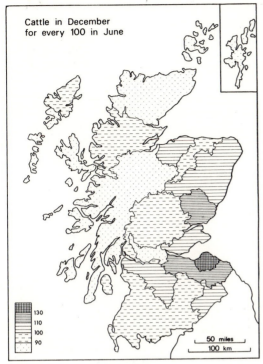

Cattle in December for every 100 in June

130
110
100
90

50 miles
100 km

Fig 132

lowlands and in the south-west, with the Merse the only major region in the lowlands where cattle accounted for fewer than 50 per cent. Orkney was again anomalous, but elsewhere in the islands, in the western Highlands and in a few areas in the Southern Uplands cattle accounted for fewer than 20 per cent of all livestock units. Fig 131 provides a parallel assessment in relation to grazing units, and shows a very similar pattern, with sheep more important than cattle only in the Southern Uplands, the Highlands and the islands, with the notable exception of Orkney.

CATTLE IN WINTER

Number in December 1965: 1,986,097

The preceding maps show the distribution and relative importance of cattle in summer; but while there are only small differences in total numbers of cattle in Scotland in June and December, there are differences in their distribution. Fig 132 shows, on a county basis, the ratio of cattle in December and June 1965. Throughout most western and northern counties there were fewer cattle in December than in June, whereas this relationship was reversed in eastern counties, notably in East Lothian, where numbers in December were 20 per cent higher. Most of these changes were due to movements of beef cattle, especially of store stock for fattening; for numbers of dairy cattle are much more stable throughout the year. Of course, there are also local movements, especially from upland to lowland, but these are largely obscured by the use of the county as the mapping unit. Nevertheless, the map will serve as a reminder that all maps of cattle, especially of beef cattle, need to be interpreted in the light of seasonal changes.

BEEF CATTLE

Beef cattle are the most important enterprise on Scottish farms, accounting for 30·4 per cent of agricultural output in 1965 and 32·5 per cent in 1972. This dominant place of beef cattle contrasts with the situation in England and Wales, for, although beef cattle are of increasing importance in all three countries, beef production

tural land. The highest densities were in Buchan and Ayrshire, but the map also brings out the importance of Galloway and the Solway lowlands. It also shows the contrasts between the Southern Uplands, where there were generally 5–15 cattle for every 100 acres, and the Highlands, with less than 5, and between the cropping districts north and south of the Forth, with the Lothians and the Merse having fewer than 25 cattle per 100 acres. Fig 129 provides a rather different perspective, relating numbers of cattle and acreages of crops and grass; the distribution reflects the lower proportion of improved land which is cropped in western districts south of the Great Glen, which have the highest densities of 60 or more cattle per 100 acres.

Cattle were the most important class of livestock as expressed in livestock units and the variations throughout Scotland in 1965 are shown in Fig 130. Cattle were relatively most important in the centre and west of the central

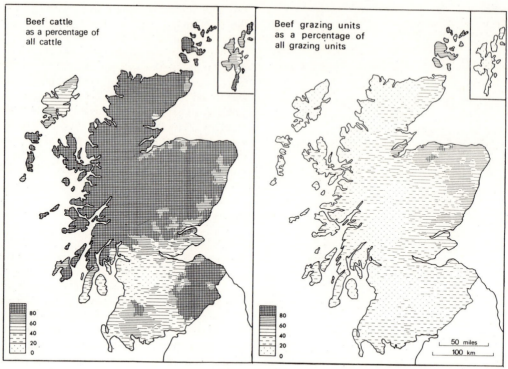

Figs 133–134

south of the Border is much more closely related to dairy farming, of which it is to a large extent a by-product, whereas beef production in Scotland is more frequently undertaken as an enterprise in its own right, and only a third of the Scottish-born calves fattened for beef come from the dairy herd. A larger urban market and a more favourable environment for the keeping of dairy cattle are major factors in explaining this contrast, as is the unsuitability for beef production of the Ayrshire, the dominant dairy breed in Scotland. For these reasons, too, a division into beef and dairy cattle can more easily be undertaken in Scotland.

BEEF CATTLE AND OTHER STOCK

Beef Cattle as a percentage of all cattle: Mean 72·8; SD 25·2; Max 100·0; Min 12·4

Beef Cattle as a percentage of all grazing units: Mean 36·2; SD 20·3; Max 91·9; Min 1·1

Although, as Figs 134 and 135 show, beef cattle are widespread, there are considerable differences in their relative importance. In 1965 they predominated in all parts of the mainland except the lowlands of central and south-west Scotland, the lowest proportions occurring in the western half of the central lowlands (Fig 133). When beef cattle are expressed as a percentage of all grazing units, there are obvious contrasts, for they yield pride of place to sheep in the uplands and compete with them on the upland margins (Fig 134). In the lowlands, too, grazing is shared with both dairy cattle and sheep, so that in the Merse, where 80 per cent or more of the cattle were beef animals, these represented less than 40 per cent of all grazing units. The North East, with 39 per cent of the beef cattle, was by far the most important area, with beef cattle accounting for 60 per cent or more of all grazing units.

BEEF CATTLE

Number: 1,343,245 in 1965; 1,746,722 in 1972
Beef cattle per 100 acres of agricultural land:
Mean 14·4; SD 11·8; Max 53·5; Min 0·2

Beef cattle have long been separately enumerated in the Scottish agricultural census and include all beef cows kept for breeding and their replacements, all male cattle other than bulls and bull calves in the dairy herd, and all other female cattle. The North East, where beef cattle accounted for 33 per cent of gross output in 1961, is the premier region and contained 40 per cent of the beef cattle in that year (although these estimates of output include transfers of store cattle only between regions and exclude Irish cattle and calves transferred from England).

Fig 135 records their distribution in 1965 and shows the importance of Buchan and the contrast between Orkney and Shetland; it also demonstrates the widespread occurrence of beef cattle throughout the lowlands and on the lower slopes of the uplands. This impression is confirmed by the fact that only in the Highlands, where there are many smallholdings, was the proportion of holdings with beef cattle less than 50 per cent (Table 16).

Fig 136 provides confirmation of this predominantly lowland distribution, for there were generally 10 or more beef animals per 100 acres throughout the lowlands, compared with under 5 in the uplands. It also confirms the prominence of the north-east, with 40 or more cattle per 100 acres in parts of Buchan, and points a contrast between the cropping districts north of the Forth and those to the south, where beef cattle were less numerous. The map also shows the importance of beef cattle in Galloway and Dumfriesshire. This contrast between upland and lowland and between west and east is further increased in winter.

This predominance of north-east Scotland in beef production is partly due to its relative disadvantage in respect of other enterprises, partly to farm structure and partly to reputation; for this is the home of the Aberdeen Angus breed. The north-east is less suited to cash crop production than areas further south, while its harsher climate and greater distance from the

TABLE 16

Beef Cattle by Regions in 1961 and 1967

High- land	North East	East Central	South East	South West	Scot- land
Percentage of gross output in 1961					
20	33	25	20	17	22
Percentage of beef cattle in 1961					
13	40	18	11	18	100
Percentage of holdings with beef cattle in 1967					
44	72	58	58	59	57

Source: Agricultural Census and *Scottish Agricultural Economics*, Vol 15

main centres of population place it at a disadvantage in milk production compared with central and south-west Scotland. On the other hand, the predominance of small farms, on which family labour can meet the irregular demands of livestock rearing, makes beef production a suitable enterprise.

Although Scotland is self-sufficient in beef, its production in Scotland cannot be considered in isolation. Some beef calves, estimated at about 60–70,000 in 1964/5, are sent to England, while about 100,000 young calves from England and Wales, mostly cross-bred or Friesians, are sent to Scotland from the dairying areas of southern England and Wales to be fattened for beef, ie a net addition of about 30–40,000 calves. A rather larger number of older store cattle, estimated at 95,000 in 1964/5, come from the principal dairying areas in Ireland. There are other complex movements within Scotland, notably of calves from dairy herds in the south-west to the east, and from uplands to lowlands.

Beef cattle are found on most types of farms, for they can play either a major or minor role in farming systems. Sales of calves and cattle in 1965 accounted for the highest proportion of gross output on upland and rearing-with-arable farms (Table 17); they are both relatively and absolutely less important on hill sheep farms, and their relative importance declines on other types of farm in competition with other enterprises.

Beef cattle

1 dot represents 200 head

50 miles

100 km

Fig 135

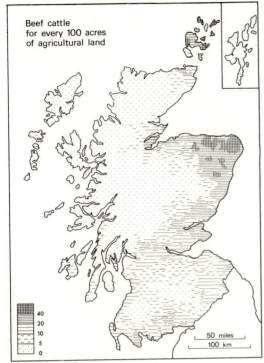

Beef cattle
for every 100 acres
of agricultural land

40
20
10
5
0

50 miles
100 km

Fig 136

TABLE 17

*Contribution of Cattle and Calves to Gross Output
in 1965 by Farm Type*

Hill sheep	Upland	Rearing with arable	Arable rearing and feeding	Crop-ping	Dairy-ing
Percentage of gross output					
18·5	43·9	45·7	37·1	21·8	15·6
Gross output from cattle and calves in £s per farm					
903	2,317	2,526	2,271	2,402	1,471
Gross output from cattle and calves in £s per acre					
0·3	1·8	8·7	14·3	8·8	7·3

Source: *Scottish Agricultural Economics*, Vol 17

In evaluating these figures some allowance must, of course, be made for differences in farm size, for the output per acre is considerably higher on the smaller, more intensively managed lowland farms.

BEEF COWS

Number: 317,425 in 1965; 448,719 in 1972
Beef cows per 100 acres of agricultural land:
Mean 3·0; SD 2·6; Max 16·8; Min 0·0
Beef cows as a percentage of all cows: Mean 59·8; SD 31·7; Max 100·0; Min 0·0

An estimated 76 per cent of all calves born in Scotland in 1964/5 and fattened for beef were born to cows in beef breeding herds, the remainder being surplus calves from the dairy herd. This disparity does not reflect differences in numbers of beef and dairy cows, for the number of beef cows (which has continued to rise since) exceeded that of dairy cattle for the first time only in 1966; it is rather that a higher proportion of female calves is required as replacements in the dairy herd and many others (estimated at 37 per cent of all dairy calves in 1962/3) are slaughtered as young calves. Beef cows were found on 37 per cent of all holdings in 1965, with the highest proportion in the North East Region, which had the largest number (Table 18). Hill cattle subsidy was paid on approximately 82 per cent of beef cows in 1965, indicating that they formed part of regular breeding herds kept on hill land, and it is therefore not surprising to find that beef cows were located mainly on the margins of the uplands, particularly on the eastern edge of the Grampians (Fig 137), though the remaining cows were widely distributed throughout Scotland; only in the western Highlands and the higher parts of the Grampians and Southern Uplands were there few beef cows.

TABLE 18

Beef Cows in 1965 and 1972

High-land	North East	East Central	South East	South West	Scot-land
Percentage of holdings with beef cows in 1965					
34	53	33	31	24	37
Percentage of beef cows in each region in 1965					
19	32	16	13	20	100
Percentage of beef cows in each region in 1972					
16	29	15	12	28	100

Source: Agricultural Censuses

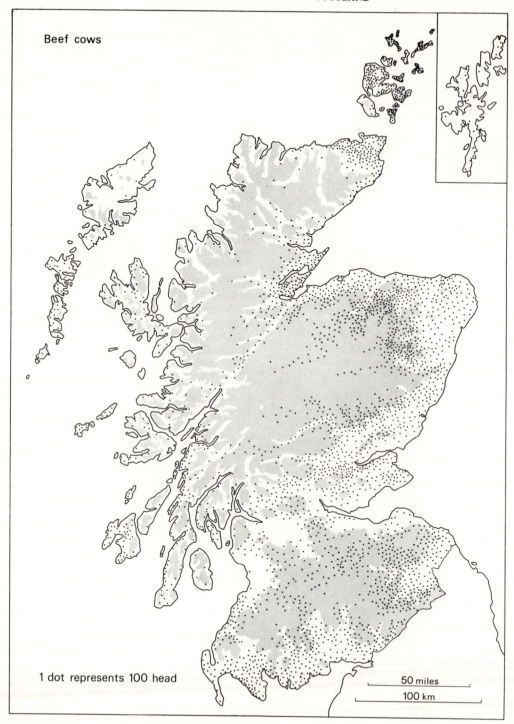

Beef cows

1 dot represents 100 head

50 miles
100 km

Fig 137

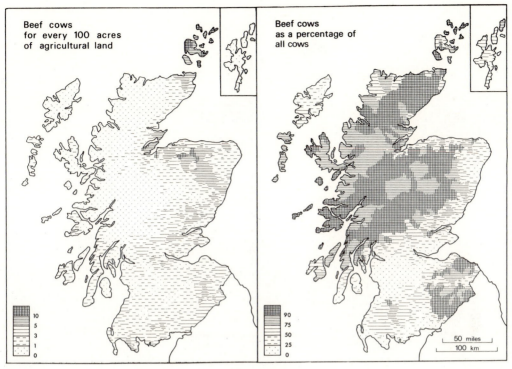

Figs 138–139

The importance of the upland margins is also revealed by Fig 138, which shows the density of beef cows in each parish in 1965. The highest values occurred on the eastern flanks of the Grampians and, to a lesser extent, in the lower Tweed basin, though the plain of Caithness and Orkney were also important areas; on the eastern lowlands, on the other hand, there were few beef cows. The map also illustrates a contrast, repeated in many maps, between the Southern Uplands and the Highlands (though owing to the large size of parishes, distributions in these areas should be interpreted with caution).

Fig 139 shows the relative importance of beef and dairy cows and resembles Fig 133 in many respects. Values were high in the Highlands (though numbers were small) and in the Tweed basin, and lowest in the western half of the central lowlands and the lowlands of the south-west; there was everywhere a contrast between the uplands and the lowlands, notably in the north-east and south-west, and most marked where dairying gives way to beef around Aberdeen.

HILL CATTLE SUBSIDY

Fig 140 shows the distribution of beef cows on which hill cattle subsidy was paid in 1967, when they numbered 304,135 (84 per cent of all cows), compared with 261,517 in 1965 and 396,000 in 1972. Owing to the high proportion of all beef cows on which hill cattle subsidy is paid,

TABLE 19

Hill Cattle Subsidy 1967

High-land	North East	East Central	South East	South West
Percentage of cows on which hill cattle subsidy was paid				
95	87	68	74	92

Source: Department of Agriculture and Fisheries for Scotland

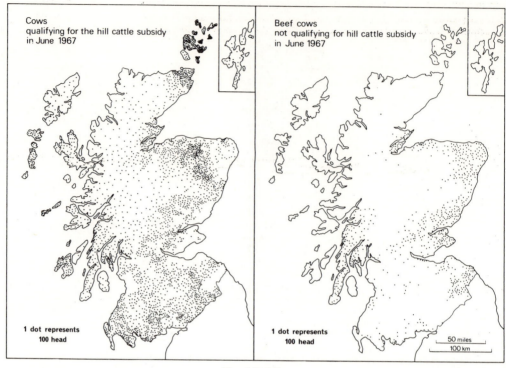

Figs 140–141

their distribution strongly resembles that of all beef cows (Fig 137); Table 19 shows the proportion in each region. A more interesting distribution is therefore that of beef cows on which hill cattle subsidy was not paid in 1967, though occupiers may have received the beef cow subsidy, introduced in 1966. These animals numbered 56,028 and were found in all lowland parishes, but were particularly prominent along the eastern lowlands from the Moray Firth to the Merse (with the exception of the Lothians) and in Argyll, especially in Kintyre (Fig 141).

Unlike hill sheep and dairy cattle, few of the beef cows are in pure-bred herds, most of which consist of either West Highland or Galloway, breeds which are most prominent in their name areas; rather more common are pure-bred animals kept for crossing on farms which also have cross-bred herds. Yet a similar zonation can be observed to that which exists in sheep farming, with an increasing admixture of lowland breeds as the quality of the land improves. Four

breeds are of particular importance, Aberdeen Angus, Beef Shorthorn, Galloway and Hereford, though there are no published data from which the distribution of the different breeds can be mapped. Most beef cows will be crosses, and most of those qualifying for the hill cattle subsidy will be first crosses, especially blue-greys (the progeny of a white Shorthorn bull and an Angus or Galloway cow). Aberdeen Angus and Hereford are the chief breeds of bulls used on cross-bred cows; other important breeds of bull are Highland and Shorthorn. On the better uplands of the north-east, where cows are not eligible for the hill cattle subsidy, second and third crosses are common, especially the former, and Aberdeen Angus and Hereford bulls are again important, with some Beef Shorthorn.

BEEF COWS AND FARM TYPE

The breeding of beef cattle is undertaken on

many kinds of farms and even 6 per cent of intensive holdings had beef cows in 1968. Proportions of holdings with beef cows were high on all the main types of livestock farms, falling below 50 per cent only on cropping and dairy farms (Table 20). As this table also shows, these proportions have increased with the rise in the number of beef cows. These figures may, however, give a misleading impression of the contributions of the different types, and Table 20 also shows the share of all beef cows on holdings of each type; it demonstrates the importance of upland and rearing with arable farms, the difference between the two sets of figures in part reflecting reclassification and in part a revision of the threshold for full-time farms.

Information is also available about the sizes of herds, which shows that, while average size of herd is increasing, the range of difference between the types is not large (Table 21). The averages should, however, be interpreted with caution, since 33 per cent of beef cows were in herds of 50 or more cows in 1964 and 44 per cent in 1968, and 8 and 15 per cent respectively in herds of 100 or more cows. It is noteworthy that the largest herds were on the less intensively managed farms where most of the beef cattle were to be found.

TABLE 20

Beef Cows and Farm Type in 1962, 1964 and 1968

Hill sheep	Upland	Rearing with arable	Rearing with intensive livestock	Arable rearing and feeding	Cropping	Dairy	All full-time	Part-time
			Percentage of all holdings with beef cows in 1964					
76	82	80	70	65	39	14	50	24
			Percentage of all holdings with beef cows in 1968					
81	88	81	65	68	35	17	52	21
			Percentage of all beef cows in 1962					
7	33	30	3	4	9	4	91	9
			Percentage of all beef cows in 1968					
10	46	17	2	4	7	4	89	11

Source: *Scottish Agricultural Economics*, Vol. 15, and *The Changing Structure of Agriculture*

TABLE 21

Beef Cows, Farm Type and Size of Holding in 1964 and 1968

Hill sheep	Upland	Rearing with arable	Rearing with intensive livestock	Arable rearing and feeding	Cropping	Dairy	All full-time	Part-time
			Average herd size in 1964					
21	33	20	7	11	16	15	20	4
			Average herd size in 1968					
29	42	23	15	14	·20	14	28	5

Source: Agricultural Censuses

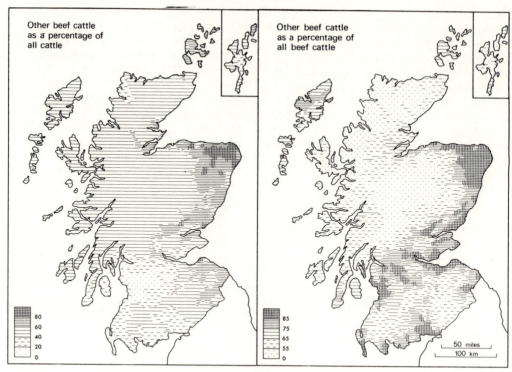

Figs 142–143

TABLE 22

Average Age of Slaughter of Fat Cattle on Full-time Farms 1964/5–1968/9

Age of slaughter (years)	Highland	North East	East Central	South East	South West	Scotland
		Percentage of all slaughterings				
under 1 year	3·8	1·1	2·4	1·4	0·8	1·5
1 year and under 2	68·8	63·8	49·4	63·5	57·0	58·4
2 years and over	27·4	35·1	48·2	35·1	42·2	40·1

Source: *Scottish Agricultural Economics*, Vol 20

TABLE 23

Other Beef Cattle in 1965 and 1972

	Highland	North East	East Central	South East	South West	Scotland
		Percentage of other beef cattle in each region				
1965	10	43	16	10	21	100
1972	9	38	16	11	26	100
		Percentage of all cattle				
1965	47	69	54	52	27	48
1972	47	67	56	56	35	50
		Percentage of all beef cattle				
1965	57	78	71	67	73	72
1972	54	74	69	66	76	69

Source: Agricultural Censuses

OTHER BEEF CATTLE

Number: 967,343 in 1965; 1,197,955 in 1972
Other beef cattle as a percentage of all cattle:
Mean 50·6; SD 17·5; Max 92·6; Min 11·3
Other beef cattle as a percentage of beef cattle:
Mean 71·6; SD 13·4; Max 100·0; Min 47·7

Other beef cattle, ie all male and female cattle other than cows, dairy herd replacements, and bulls and bull calves, represent the various stages of breeding, rearing (in so far as this is separately identifiable) and fattening; they also include replacements for the beef breeding herd, which account for about a tenth of all other beef cattle. The age groups used in the census are not, however, wholly satisfactory as a basis for distinguishing the different stages of beef production. Of particular significance in this respect is the trend towards the slaughter of cattle at younger ages; for example, the average age at slaughter fell from 24·3 months in 1964/5 to 22·8 in 1968/9 and the proportion slaughtered at ages of two years and over fell from 54·6 per cent to 30·7 per cent (though the latter figure ranged from 27·4 per cent in the Highland Region to 48·2 per cent in the East Central Region). On the other hand, the number of calves reared intensively and slaughtered at ages under one year is small and did not exceed 3 per cent throughout the period; even calves reared semi-intensively (ie slaughtered at ages between twelve and eighteen months) did not exceed 10 per cent. Thus, while the great majority of fat cattle are slaughtered at ages exceeding one year, the division between cattle which are one-year old and under two and those which are two-years old and over does not separate those being reared from those being fattened; indeed,

in every region, more were slaughtered before reaching the age of two years than after (Table 22). About three-quarters of the cattle slaughtered were of beef breeds, though the proportion varied from 90 per cent in the East Central Region to 43 per cent in the South West, and also with age of slaughter; dairy breeds were particularly prominent among those purchased as young calves at under two-months old and also had an above average share of the Scottish-bred cattle purchased at ages of one-year old and over.

These considerations do not, of course, affect the relative importance of other beef cattle as a whole. Figs 142 and 143 show such cattle in relation to numbers of all cattle and all beef cattle respectively. Both illustrate the importance of north-east Scotland for the rearing and fattening of beef cattle. Table 23 provides a general view and shows the stability of the general pattern despite the considerable increase in the number of beef cattle. In Fig 142, the lowest values were understandably to be found in the main dairying areas in the western half of the central lowlands and in south-west Scotland, whereas in Fig 143 high values occur throughout the lowlands and low values in the uplands and in most of the islands (where numbers were in any case small). Variations between the various age groups of other beef cattle are examined in Figs 144 to 155.

Although, as the subsequent discussion of the three age groups of other beef cattle reveals, there are some differences in the type of farm on which each is most important, information on the association of all other beef cattle with farm type shows the importance of upland, rearing with arable and cropping farms, each of which plays a major role in the progression from breeding to fattening (Table 24).

TABLE 24

Other Beef Cattle and Farm Type in 1965

Hill sheep	Upland	Rearing with arable	Rearing with intensive livestock	Arable rearing and feeding	Cropping	Dairy	Intensive	All full-time
			Percentage of other beef cattle by farm type					
3	18	25	5	11	17	13	1	93

Source: Agricultural Censuses

OTHER BEEF CATTLE UNDER ONE-YEAR OLD

Number: 471,250 in 1965; 617,955 in 1972

Other beef cattle under one-year old as a percentage of all beef cattle: Mean 36·9; SD 8·0; Max 77·5; Min 7·6

Other beef cattle under one-year old as a percentage of other beef cattle: Mean 54·2; SD 17·4; Max 90·3; Min 8·2

Other beef cattle under one-year old comprised mainly steer calves and those heifer calves not required for herd replacements in the dairy herd. They were mainly cattle being bred or reared for beef, though they might also include some of the dairy calves sold within a few weeks for slaughter as bobby calves and a small number intended for fattening under intensive systems and slaughter before the age of one; they will also have included, as the discussion on Fig 147 shows, some dairy calves from England and Wales. Of this total, approximately two-fifths were female. Such young cattle were well distributed throughout the lowlands and upland margins of Scotland in 1965, being few only in the higher parts of the Southern Uplands and in the Highlands, and in the crofting districts of the mainland and, with the exception of Orkney, the islands (Fig 144). They were, however, most numerous in the North East Region, which had 37 per cent of the Scottish total in 1965 and 31 per cent in 1972. Data are not available on the proportion of holdings with other beef cattle under one as a whole, but evidence on other beef cattle under six-months old in 1967 shows that the North East also had the highest proportion (Table 25). On the other hand, because of the importance of older beef animals in this region, beef cattle there

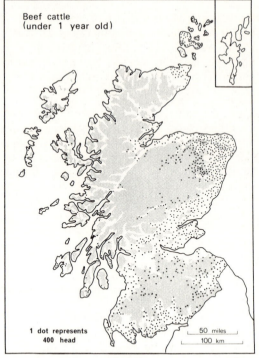

Fig 144

under one-year old accounted for the smallest proportion of all other beef cattle.

Fig 145 shows beef calves under one as a percentage of all beef cattle. The variations between parishes were not large and percentages were generally highest in the uplands, where numbers were small and the emphasis wholly on breeding and rearing. When the basis of comparison is altered to 'other beef cattle' the relative importance of the uplands is even more marked, with young beef cattle accounting for 60 or more per cent throughout the uplands, and with the lowest percentages in the east, especially in the north-east, where the contrast between upland and lowland was particularly marked (Fig 146). Such young beef cattle were associated with a wide range of farm types, but it is clear from Table 26 that upland farms, where many of the breeding cows were to be found, were the most important type. The average number of young beef cattle on each holding was also highest on holdings of this type, with an average of 44 animals, compared with 29 on rearing with arable farms and 21 on hill sheep farms. This

TABLE 25

Other Beef Cattle Under One-year Old in 1965 and 1972

	High- land	North East	East Central	South East	South West	Scot- land
Percentage of other beef cattle under one-year old in each region						
1965	13	37	15	11	24	100
1972	11	31	15	12	30	100
Percentage of holdings with other beef cattle under six-months old						
1967	32	53	36	37	39	40

Source: Agricultural Censuses

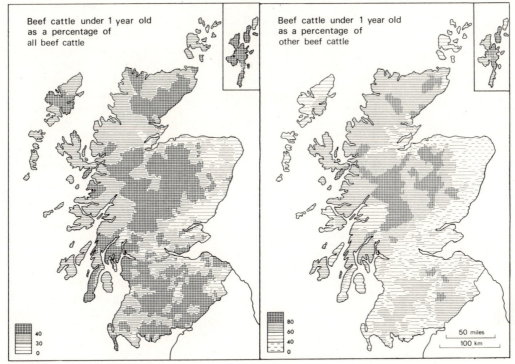

Figs 145–146

relationship was true for both male and female cattle, though male cattle were relatively more important on cropping and dairy farms. Apart from the reallocation between upland and rearing with arable farms discussed in Section 7, there is no reason to believe that the pattern would have changed greatly since 1965.

MOVEMENTS OF YOUNG CATTLE

Further light can be thrown on the distribution of young beef cattle and on the mingling of the various streams of cattle intended for fattening and slaughter for beef by examining both the relationship between young cattle and breeding stock and the sales and purchases of calves. For there are differences in the proportion of steers born in each parish which are retained for rearing for beef, and also transfers of such animals between holdings in Scotland and into and out of Scotland.

TABLE 26

Other Beef Cattle Under One-year Old and Farm Type in 1970

Hill sheep	Upland	Rearing with arable	Rearing with intensive livestock	Arable rearing and feeding	Cropping	Dairy	Intensive	All full-time

Percentage of other beef cattle under one-year old on each type

Hill sheep	Upland	Rearing with arable	Rearing with intensive livestock	Arable rearing and feeding	Cropping	Dairy	Intensive	All full-time
5	40	17	2	4	10	12	½	89

Source: Agricultural Census

Steers under 1 year old
as a percentage of
all cows

100
60
40
20
0

50 miles
100 km

Fig 147

Ratio of Steers and Cows

Steers under one-year old per 100 cows: Mean
49; Max 378; Min 6

Fig 147 shows the proportion of steers under
one-year old, which form part of the beef herd,
and all cows, whether in dairy or beef herds,
since both are a source of beef animals. If there
were no differences in calving rates and losses
throughout Scotland, no transfers of steers
before they were one-year old and no changes in
numbers of cows, the ratio would be very
similar throughout the country. In fact, it varies
greatly, with generally higher values in the east
than in the west.

The localisation of much of the dairy herd in
the western half of the central lowlands and in
south-west Scotland and the low ratio of beef
to dairy cows in these areas are partly responsible
for the lower ratio of steers to cows. The main
breed of dairy cow, the Ayrshire (Fig 161) has
not been regarded as a suitable type for beef
and most steer calves were formerly sold slink,

ie within a few weeks of birth for slaughter.
Although the proportion has declined since 1941,
when the calf subsidy was introduced to encourage the retention of calves for beef, 24 per
cent of the calves from the dairy herd were
disposed of in this way in 1966 (Table 27), and
possibly a fifth of the steer calves.

A more important reason is the transfer of
young steers to other farms, often in different
parts of the country. Most of the dairy calves
intended for beef are transferred to other farms
and a small proportion moves to eastern
Scotland. Other dairy calves, intended mainly
for fattening, come from herds in England and
Wales and, either directly or after spending some
time on another farm, move to eastern Scotland,
particularly to the rearing farms which are
characteristic of the north-east. Some of the
steers under one-year old in the Scottish beef
herd are also sold at ages below one year and
many of these, too, move to farms in eastern
regions, particularly from the Highland Region
(especially the crofting districts) to the North
East.

Similar movements take place among heifer
calves, but numbers of these are additionally
affected by the different proportions of heifer
calves required for herd replacements, since
herd life is some two-thirds longer in beef herds
than in dairy herds and the proportion of young
female cattle required as herd replacements
correspondingly smaller. Approximately two-
thirds of the female dairy calves born in Scotland
are required as herd replacements and many of
the remainder are sold for slaughter as bobby
calves; by contrast, only about two-fifths of the
beef heifers are required as replacements and the
remainder are reared for beef. Separate figures
are not available for steers and heifers, but it was
estimated by W. J. Carlyle in 1966 that, of the
319,000 calves born in dairy herds in 1966,
133,000 were retained for beef and 92,000 of
these were transferred to other farms, all but
18,700 of them within the same region (Table 27).
Of the 299,000 beef calves, 239,000 were available for rearing for beef and 180,000 of these
were transferred to other farms, 110,000 of them
as calves under one-year old. In addition to the
movements into and within Scotland, relatively
small numbers of suckled calves are also sold to

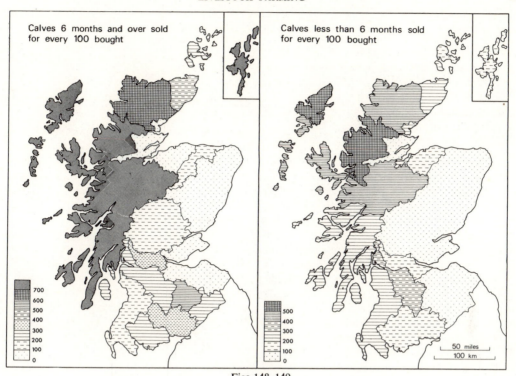

Figs 148–149

TABLE 27

Calves Intended for Beef in 1966

	Highland*	North East	East Central	South East	South West	Scotland
	Beef herd in thousands					
Total calves born	88	68	50	33	60	299
Total for feeding	64	55	39	28	53	239
	Dairy herd in thousands					
Total calves born	20	33	33	19	215	319
Sold slink	3	3	7	4	61	78
Total for feeding	11	22	16	9	75	133
	Transfers in thousands					
Net, calves 6 months†	−2	+11	+3	+3	−15	—
Net, calves 6–12 months†	−22	+18	+2	−7	−9	−18
From England and Wales	1	44	7	13	13	78
Stores for feeding at 1 year	52	151	67	46	117	433

Source: W. J. Carlyle   *includes Orkney and Caithness   † − =loss,   + =gain

England, mainly from the South East and South West Regions. Table 27 shows that the North East was a large importer of calves for rearing for beef, and the Highlands and the South West were large exporters.

CALVES BOUGHT AND SOLD IN 1965

Another view of such movements is recorded in Figs 148 and 149, which show in which counties more calves were bought than sold in the six months ending on 3 June 1965. Figs 148 and 149, show that more calves were bought in the eastern counties than were sold, whereas the Highlands and the islands were net suppliers, the latter mainly from the dairy herd and the former providing young beef animals. Both maps and the figures for movements in Table 27 record net movements; it should be noted, particularly in respect of calves of six months and over, that there are movements in both directions. On these maps, the eastern parishes of Inverness, Ross and Sutherland, which have strong affinities with eastern counties, have been shown separately from the remainder of those counties; the blurring of east/west contrasts, which is often characteristic of maps of county data, is thus avoided.

OTHER BEEF CATTLE ONE-YEAR OLD AND UNDER
TWO

Number: 372,977 in 1965; 473,480 in 1972
Other beef cattle one-year old and under two as
    a percentage of all beef cattle: Mean 23·7;
    SD 12·9; Max 58·6; Min 0·4
Other beef cattle one-year old and under two as
    a percentage of other beef cattle: Mean 31·2;
    SD 13·1; Max 66·7; Min 0·8

The distribution of other beef cattle which
were one-year old and under two in June 1965
was broadly similar to that of younger beef
cattle (Fig 144), but more localised (Fig 150), for
there were few older animals in the uplands and
the concentration of beef cattle in Buchan and
surrounding parishes was even more marked;
indeed, 29 per cent of such cattle were in
Aberdeenshire alone and this dominance has
not changed greatly since (Table 28). Some 41
per cent of holdings had other beef cattle in this
age group in 1967, with the highest proportion
in the North East; the very large numbers of
small holdings in the Highlands help to account
for the low percentage in that region. Some 59
per cent of all cattle in this group were male
(61 per cent in 1972), though the proportion
varied greatly, from 76 per cent in East Lothian
to 34 per cent in Sutherland.

    Considered as a proportion of all beef cattle,
other beef cattle one-year old and under two
differed from younger beef cattle (Fig 145) in
that the highest proportions were recorded in the
lowlands; nevertheless, north-east Scotland was
again the main area, with cattle in this group
accounting for 30 per cent or more of all cattle,
compared with under 10 per cent in the Highlands (Fig 151). The distribution of other beef
cattle in their second year as a percentage of all
other beef cattle was very similar, with values

Beef cattle
(1 to 2 years old)

1 dot represents
400 head

50 miles
100 km

Fig 150

above 40 per cent throughout the north-east and
high values scattered elsewhere throughout the
lowlands and in Orkney (Fig 152).

    Other beef cattle one-year old and under two
were found on all types of farm, but were least
numerous on hill sheep farms; unlike younger
beef cattle, which were most strongly represented on upland and rearing with arable farms,
large numbers of these cattle were also found on
cropping farms (Table 29). In part because they
accounted for 78 per cent of all other beef cattle
of one-year old and over, the latter had a very
similar relationship to farm type.

TABLE 28
Other Beef Cattle One-year Old and Under Two in 1965, 1967 and 1972

	Highland	North East	East Central	South East	South West	Scotland
Percentage of other beef cattle one-year old and under two in each region						
1965	7	48	16	9	19	100
1972	7	45	16	10	23	100
Percentage of holdings with other beef cattle one-year old and under two						
1967	25	58	41	42	44	41

Source: Agricultural Censuses

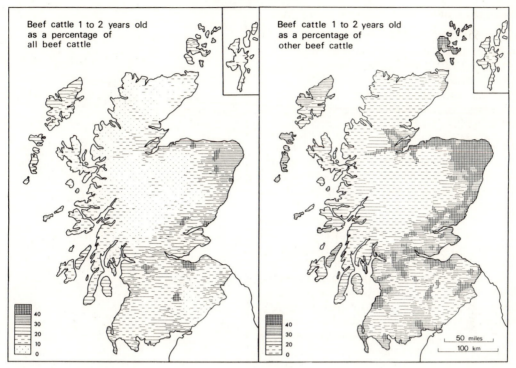

Beef cattle 1 to 2 years old
as a percentage of
all beef cattle

Beef cattle 1 to 2 years old
as a percentage of
other beef cattle

Figs 151–152

TABLE 29

Other Beef Cattle One-year Old and Under Two and One-year Old and Over, and Farm Type in 1968

Hill sheep	Upland	Rearing with arable	Rearing with intensive livestock	Arable rearing and feeding	Cropping	Dairy	Intensive	All full-time
Percentage of other beef cattle one-year old and under two in 1968								
2	20	23	4	10	22	9	1	89
Percentage of other beef cattle one-year old and over in 1968								
2	19	23	3	10	23	8	1	90

Source: Agricultural Census

OTHER BEEF CATTLE TWO-YEARS OLD AND OVER

Number: 123,116 in 1965; 106,520 in 1972

Other beef cattle two-years old and over as a percentage of all beef cattle: Mean 7·6; SD 7·8; Max 53·5; Min 0·0

Other beef cattle two-years old and over as a percentage of other beef cattle: Mean 9·5; SD 8·7; Max 55·1; Min 0·0

Other beef cattle two-years old and over, representing both advanced stores and cattle being fattened for slaughter, were once the main source of beef, but are now of declining importance in both absolute and relative terms, and accounted for less than half the steers and heifers slaughtered for beef in 1965/6. About three-quarters of these cattle were male in 1965 and 1972, though the proportions ranged widely, being lowest in the Highlands, with 43 per cent in Sutherland, and highest in the East Central and South East Regions, with 91 per cent in Berwickshire. Numbers vary somewhat between summer and winter; the June total in 1972 was 91 per cent of that in December and totals in the East Central and South East Regions were 78 and 67 per cent respectively, reflecting the importance of yard feeding in winter in those areas.

The distribution of other beef cattle two-years old and over is very similar to that of other beef cattle one-year old and under two, though there has been some tendency for the importance of the North East to increase rather than diminish (Table 30 and Fig 153). They were much less common on holdings than were other beef cattle,

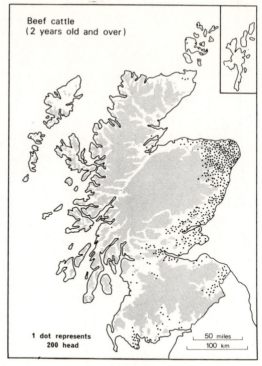

Beef cattle (2 years old and over)

1 dot represents 200 head

50 miles
100 km

Fig 153

although the North East was again the region with the highest proportion. Even these figures may be somewhat misleading, for the final stages of fattening are strongly associated with large farms; in 1961 farms selling 100 or more cattle, accounting for 8 per cent of the full-time holdings selling any fat cattle, provided 46 per cent of all the cattle sold. In relative terms the importance of the north-east was less marked, parishes in which such cattle accounted for 20 per cent or more of all beef cattle being scattered throughout the eastern lowlands (Fig 154); such cattle were also relatively most numerous in the Outer Hebrides. The map showing these older beef cattle as a proportion of other beef cattle is very similar (Fig 155).

Other beef cattle two-years old and over were most important on rearing with arable farms and on cropping farms in 1968, the first year for which published information on this aspect of their distribution is available; these proportions do not differ greatly from those for all other beef cattle one-year old and over.

TABLE 30

Other Beef Cattle Two-years Old and Over in 1965, 1967 and 1972

	High-land	North East	East Central	South East	South West	Scot-land
Percentage of other beef cattle two-years old and over in each region						
1965	6	46	21	9	18	100
1972	4	50	19	7	20	100
Percentage of holdings with other beef cattle two-years old and over						
1967	8	22	13	14	16	15

Source: Agricultural Censuses

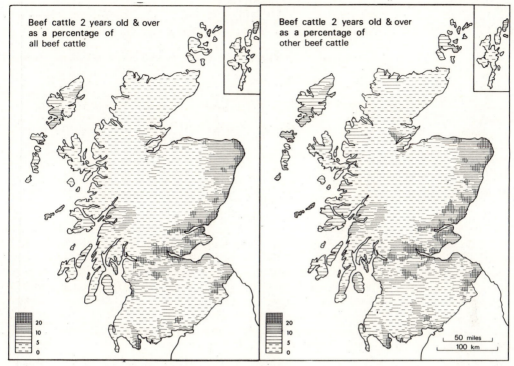

Figs 154–155

TABLE 31

Other Beef Cattle Two-years Old and Over and Farm Type in 1968

		Rearing with arable	Rearing with intensive livestock	Arable rearing and feeding	Cropping	Dairy	Intensive	All full-time
Hill sheep	Upland							
		Percentage of other beef cattle two-years old and over						
1	18	25	2	11	26	7	—	92

Source: *Agricultural Statistics*

CATTLE FATTENING

The division of other beef cattle one-year old and over into those under two and older cattle is unsatisfactory in that this boundary has no particular functional significance. Beef cattle are slaughtered at all ages covered by these groups, and other beef cattle between one- and two-years old include both store cattle and animals being fattened for slaughter; they also include some female cattle intended as replacements in the beef herd and so are roughly equi-valent to the number of first-line replacements, viz 43,233 in-calf heifers in 1965 (80,781 in 1972). Furthermore, while additional transfers take place between regions in Scotland and between Scotland and other parts of the British Isles, available information does not distinguish those cattle under two-years old from older cattle. Table 32 shows the net movements between regions of beef cattle of one-year old and over. Changes within Scotland are much more sub-

TABLE 32

Transfers of Store Cattle Over One-year Old in 1966

	Highland*	North East	East Central	South East	South West	Scotland
			In thousands			
Stores for feeding at 1-year old	52	151	67	46	117	433
Net transfers	−40	+24	+9	−2	−24	−33
Irish stores for feeding	1	26	30	18	5	80
Total for feeding	13	201	106	62	98	477

*includes Orkney and Caithness
− = net loss,   + = net gain

Source: W. J. Carlyle

stantial, for many transfers take place between farms in the same region; but even regional movements include both gains and losses from the same region. There were again large substantial net movements from the Highlands and the South West to eastern regions, especially the North East. The main external gain is the import of Irish stores for feeding, the main loss the transfer of some 21,000 stores to England. A different view is provided by Table 33, which records the main sources of animals slaughtered for beef in 1963/4 and compares the numbers of cattle one-year old and over with those of older store cattle intended for fattening and with those cattle qualifying for the fatstock guarantee in 1965/6. These figures must be borne in mind in this and the following commentary on other beef cattle in these two age groups.

TABLE 33

Older Store Cattle and Other Beef Cattle for Slaughter in 1963/4, 1965/6 and 1966

	Highland	North East	East Central	South East	South West	Scotland
	Percentage of cattle slaughtered in 1963/4					
Born and bred on same farm	43	19	19	29	34	24
From Scottish farms at ages under 1 year	58	39	27	19	18	28
From Scottish farms at ages of 1 year and over		29	25	19	41	30
From Ireland	—	9	29	28	5	15
From England and Wales	—	4	1	6	3	3
	Percentage of cattle one-year old and over in 1965					
	7	48	17	9	19	100
	Percentage of older stock available for feeding in 1966					
	12	48	11	6	23	100
	Percentage of cattle qualifying for fatstock guarantee in 1965/6					
	3	46	20	11	20	100

Sources: *Scottish Agricultural Economics*, Vol 15, W. J. Carlyle and Agricultural Census

TABLE 34

Systems of Feeding and Breeds of Beef Cattle in 1963/4

	Highland	North East	East Central	South East	South West	Scotland
	Percentage by different systems in each region					
Intensive	2	4	2	8	3	4
Court fed and finished ⎫		17	24	47	20	24
Grass fed and finished ⎬	98	13	18	8	37	18
Combined systems ⎭		67	56	37	41	54
All	100	100	100	100	100	100
	Percentage of beef breeds					
	80	82	90	78	43	75

Source: *Scottish Agricultural Economics,* Vol 15

As well as the many different sources from which cattle for fattening are drawn (Table 33), there are many systems of fattening cattle, which are related primarily to the supply of store cattle for fattening and the character of the farm on which they are fed, particularly its size and the availability of arable crops and crop residues. Published data are scarce, but a survey has shown that several systems could be identified in 1963/4. Intensive systems contributed only a small part of the total number of cattle fattened and were relatively most important in the South East Region (Table 34). Many of these cattle were of dairy breeds, either from Scotland or from England and Wales, and Friesians and Friesian crosses were particularly common; for beef breeds tend to put on fat too quickly. Systems of fattening on grass were more characteristic of cattle purchased at ages of one-year and upwards and such animals tended to be slaughtered at greater ages than those fattened in yards. Fattening in yards is particularly characteristic of arable farming areas where there are crop residues and where the manure made by cattle may be as important as the cattle themselves. Differences also exist between breeds, more beef breeds being fattened in yards than on grass and more cattle of dairy breeds being fattened on grass than in yards. Nevertheless, most cattle were fattened in combined sys-tems, being either court fed and finished on grass or vice versa. Quite marked regional differences also occur in the relative importance of different systems, with intensive systems relatively common in the South East, which also had the highest proportion of cattle fed and finished in yards. Grass fattening was most common in the South West, a not surprising feature in view of the fact that about 57 per cent of the animals fattened in that region were of dairy breeds. It must, however, be appreciated that these are percentage values; proportionately more cattle were finished on grass in the South West than in the North East, but the number in the North East was larger.

The emphasis in this section has been on the rearing and fattening of clean cattle, ie steers and heifers, but it must also be remembered that there are other sources of beef. The chief of these comprises cast cows and bulls from the breeding herds, especially the dairy herd, and these totalled some 124,000 in 1963/4. Such animals are either sold as soon as they have been culled or experience only a short period of fattening, so that some may be included in those returned as other cattle two-years old and over. Other beef supplies come from fat cattle imported from Ireland, which similarly spend little, if any, time on British farms.

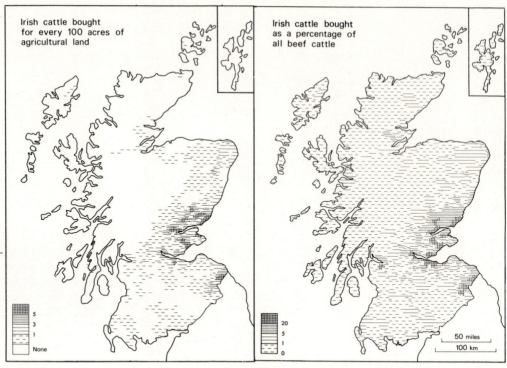

Figs 156–157

IRISH CATTLE

Irish cattle per 100 acres of agricultural land:
Mean 0·6; SD 1·2; Max 6·9; Min 0·0
Irish cattle as a percentage of all beef cattle:
Mean 3·2; SD 6·4; Max 43·7; Min 0·0

Irish cattle have already been included in the various categories of beef cattle considered earlier in this section and are the only class of livestock to be identified by origin. They consisted mainly of forward stores, those coming from the Irish Republic requiring a minimum stay of two months to become eligible for fatstock guarantees; but breeding stock have been making an increasing contribution, encouraged by the expansion of the Scottish beef industry, and accounted for about 12 per cent of Irish cattle in 1966. Most of these imports came from the Irish Republic but about a quarter (27 per cent in 1969) came from Northern Ireland, which was formerly a much more important source before the introduction of guaranteed prices for fatstock began to encourage Ulster farmers to

TABLE 35

Irish Cattle in 1966

	High-land	North East	East Central	South East	South West	Scot-land
Percentage of Irish cattle for feeding						
	100	92	90	78	67	88
Percentage of Irish cattle for feeding in each region						
	—	32	42	22	4	100
Percentage of Irish cattle for breeding in each region						
	2	21	33	11	33	100
Percentage of all Irish cattle in each region						
	—	30	41	20	9	100

Source: W. J. Carlyle

retain stock for fattening. Such animals enter Scotland almost entirely through the port of Glasgow and go mainly to farms in eastern Scotland on completion of their period of quarantine, either directly or through auction markets. The returns on which Figs 156 and 157 are based are derived from responses to a question in the June census about the number of Irish cattle purchased in the six months ending on 31 May and therefore includes both stores and animals for breeding; a similar question is asked in the December census and the distribution revealed by this is broadly similar though figures for the first half of the year understate the importance of the eastern arable regions, where most of the cattle are purchased in the autumn. Irish cattle for breeding go mainly to the North East, the East Central and South West Regions, and are relatively most important in the last; those for feeding go to the North East, East Central and South East Regions (Table 35).

The distribution of those Irish cattle recorded in June 1965 is shown in Fig 156. They were mainly of beef breeds and most numerous in the arable areas south of the Highlands where many are fattened in yards, particularly in the potato-growing areas north of the Tay. Few animals went to the Moray Firth lowlands and to the south-west because these areas had good supplies of home-bred stores. The map of Irish cattle as a percentage of all beef cattle shows a similar emphasis on the lowlands between Berwick and Stonehaven (Fig 157).

Irish cattle have shown a downward trend over the past forty years and their chief role in recent years has been as a reserve to meet deficiencies in home-bred supplies of cattle for fattening and, increasingly, as a source of breeding stock.

DAIRY CATTLE

Dairying is a less important branch of farming in Scotland than it is in other parts of Great Britain, largely because of the much smaller area of Scotland which is well suited to dairy farming and because of the smaller population. Furthermore, since the creation of marketing boards, Scottish milk has not normally been sold south of the Border, as it was formerly, while the Scots themselves drink less milk per head, consumption being only 89 per cent of that in England and Wales in 1965/6. This lesser demand is also reflected in the larger proportion of Scottish

milk which is manufactured, particularly in the Aberdeen and District Milk Marketing Board's area, which is more remote from the main centres of population than the principal dairying areas, and in the fact that the gallonage of milk which attracts government price guarantees (the standard quantity) represents a larger share of the total production than in England and Wales. Nevertheless, milk is the second most important source of farm income, accounting for 28 per cent of gross farm output in 1965 and 20 per cent in 1970. Dairy herds also contribute to Scottish beef production, providing about a third of the Scottish-born calves reared for beef as well as most of the bulls and cows slaughtered; they also provided some 78,000 bobby calves in 1966.

Dairying is especially a feature of South West Scotland, which had 69 per cent of all dairy cattle in both 1965 and 1972 (Table 36). The

TABLE 36

Dairy Cattle in 1965 and 1972

	High-land	North East	East Central	South East	South West	Scot-land
Percentage of all dairy cattle in each region						
1965	5	10	10	6	69	100
1972	5	11	10	5	69	100
Percentage of dairy cows in each region						
1965	6	12	11	6	66	100
1972	5	13	10	5	67	100
Percentage of other dairy cattle in each region						
1965	4	7	10	6	73	100
1972	4	10	10	5	71	100

Source: Agricultural Censuses

proportion of dairy cows was rather less, in part because of the widespread distribution of small non-commercial herds providing farm milk, but chiefly because the South West also provides herd replacements for other regions, a fact reflected in the higher percentage of other dairy cattle in that region.

Holdings with dairy cattle were also relatively more numerous in the South West (Table 37), though the predominance of that region was less marked, largely because of numerous non-

TABLE 37

Dairy Cattle, Dairy Cows and Other Dairy Cattle in 1967

High-land	North East	East Central	South East	South West	Scot-land
Percentage of holdings with dairy cattle					
15	14	24	28	56	25
Percentage of holdings with dairy cows					
15	13	22	24	49	23
Percentage of holdings with other dairy cattle					
7	5	14	17	50	17

Source: Agricultural Censuses

commercial herds in other regions, an interpretation which is confirmed by the higher proportions of holdings on which there were dairy cows.

A distinctive feature of the Scottish dairy industry is the large average size both of farms producing milk for sale and of the milking herds on such farms. Information on this aspect of Scottish dairying is derived from the periodic censuses conducted by the three milk marketing boards which are responsible for controlling the sale of milk from Scottish farms, viz the Scottish Milk Marketing Board, by far the largest, with jurisdiction over southern and central Scotland, ie the main dairying areas; the Aberdeen and District Milk Marketing Board, with responsibility for north-east Scotland; and the North of Scotland Milk Marketing Board, covering the remainder of the mainland. Most of the islands, with the notable exceptions of Arran, Islay and Orkney, do not come under any board. In 1965, the average size of milk-selling farms recorded in the census of that year was 167 acres (68ha) of crops and grass and 236 acres (96ha) of total area; by 1972, these figures had risen to 185 acres (75ha) and 255 acres (103ha) respectively. Over the same period the average herd increased in size from 45 cows to 62 cows. These averages are somewhat misleading because the statistical distribution of herds by size of herd is not normal, so that in 1965, for example, the median herd size was 40 cows; an increasingly large part of the cow population is thus to be found in a relatively small number of large herds, and in 1972 more than half the dairy cows were recorded in herds of 70 or more cows. While increases in average size are in part a statistical illusion, resulting from the decline in the number of herds, especially those in the smallest size groups, in part they are due to increases in the number of cows in each herd (though the existence of solidly built byres in the main dairying areas does place some restraint on increasing the size of herd).

'It was family labour, the Ayrshire cow and the Clyde valley consumer that made these farms what they are today' wrote J. A. Gilchrist in 1958 in a commentary on the dairy industry of west-central Scotland. The existence of a large and rapidly growing urban market certainly played a major part in the evolution of the Scottish dairy industry, but its location is now of much less importance, partly because of the control over marketing and transport costs exercised by the milk marketing boards and partly because of increasing efficiency in the collection of farm milk, which is now undertaken mainly by bulk tanker. Transport costs from farms to first destination are stepped, with uniform rates within zones of specified radii, and taper with distance; in the Scottish Milk Marketing Board area, for example, the rate in 1965 was 0·625d per gallon for up to 5 miles, 0·875d for between 5 and 10 miles and 1·125d for between 10 and 30 miles. In this board's area, a further degree of artificial proximity was given by nominating as first destinations a number of small towns in eastern Scotland, to avoid harming producers who might otherwise have been adversely affected by the milk marketing schemes. As a result of the declining importance of proximity to markets, physical characteristics, in respect of both suitability for dairying and the comparative advantage of milk over other enterprises, are now the major factors in the distribution of dairy farming in Scotland.

By comparison with beef animals, dairy cattle are both more sensitive to climatic conditions, especially temperature and exposure (though the Ayrshire is a relatively hardy dairy breed), and more demanding in respect of quality of feed. Dairying is thus strongly associated with the main grassland areas, which have a moderate, well-distributed rainfall and, at least along the coasts of south-west Scotland, a longer grazing season. Dairying is primarily a lowland activity, and four-fifths of the farms selling milk in 1969 had farm buildings below the 500ft (150m) contour. With increasing altitude the length of the growing season tends to decrease and, other things being equal, the costs of milk production tend to rise proportionately. In a survey conducted in 1969 by the milk marketing boards, 42 per cent of the farms above 500ft had a growing season of less than 5 months and 7 per cent a growing season of 6 or more months, compared with 10 and 40 per cent respectively on farms below 200ft (60m).

Elevation and climate are not, however, the only factors affecting the length of the growing season, which also tends to vary with soil type, being shorter on heavy soils; in the same survey, 31 per cent of farms on mostly heavy land had a grazing season of less than 5 months, compared with 17 per cent on farms on mainly light land. The relationships are by no means simple, but there is little doubt that occupiers of farms on heavy land in areas of moderate to heavy rainfall have less choice in the range of enterprises they can undertake. In general, the main dairying areas are less well suited on climatic grounds to crop production and have the further disadvantage of large areas of cold heavy land; they are best suited to summer production of milk and, partly owing to the higher proportion of milk manufactured (42 per cent in 1965/6) a higher percentage of Scottish milk (57 per cent in 1965/6) is produced in the summer six months than in England and Wales.

Dairying in Scotland is also less strongly associated with farm size than in England and Wales, where dairy farms are small both by area and by number of cows. The decline in the numbers of dairy cattle and of milk-selling farms, which has been a characteristic feature of Scottish dairying since the early 1950s, has eliminated many of the smaller herds, and farm size does not appear to be an important factor in explaining the distribution of dairy farms in Scotland. Those giving up dairying have been doing so primarily for economic reasons, though this term has been rather widely interpreted to include the better return from other enterprises, as well as problems over finance and labour supply. The latter has been especially important in the central lowlands, though family labour makes a major contribution to labour supply on dairy farms in the west.

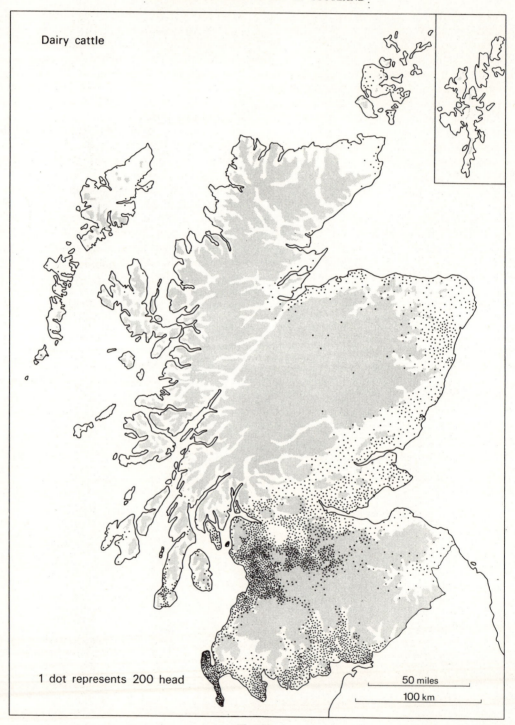

Dairy cattle

1 dot represents 200 head

50 miles

100 km

Fig 158

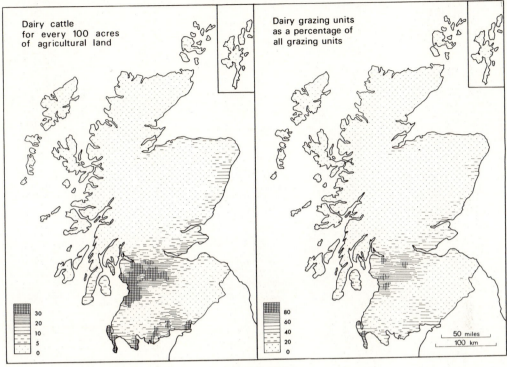

Figs 159–160

DAIRY CATTLE

Number: 691,973 in 1965; 640,933 in 1972
Dairy cattle per 100 acres of agricultural land:
  Mean 8·0; SD 11·4; Max 62·3; Min 0·0
Dairy grazing units as a percentage of all grazing
  units: Mean 18·3; SD 21·2; Max 83·4; Min 0·0

As Table 36 has shown and as Fig 158 confirms, the distinctive feature of the Scottish dairy industry is its concentration in south-west Scotland. In 1965 dairy cattle were most prominent in the lowlands of Ayrshire, in the Clyde valley and in the coastal lowlands of Wigtownshire (especially the Rhinns of Galloway) and the Solway lowlands generally. They were also fairly numerous in the eastern half of the central lowlands, in Fife and West Lothian, and in the hinterland of Aberdeen. Dairy cattle occurred in small numbers throughout the eastern lowlands; they were also a characteristic feature of the lowlands of Kintyre and of the islands in Argyll, but were of very minor importance in the remaining islands (except Orkney).

Fig 159, in which this distribution is quantified, confirms the pre-eminence of the western half of the central lowlands and the general lowland nature of this distribution; it also demonstrates the unimportance of dairy cattle in the Tweed valley.

Fig 160 provides one measure of the relative importance of dairy cattle in respect of all grazing livestock, viz sheep and cattle. It, too, shows that the western half of the central lowlands was the most important area, with dairy cattle accounting for 60 per cent or more of all grazing units. By contrast, in only a few localities in the eastern lowlands did the proportion exceed 20 per cent. Another view of the relative importance of dairy cattle is provided by Fig 162, which shows the ratio of dairy cattle to beef cattle, their main competition in the lowlands.

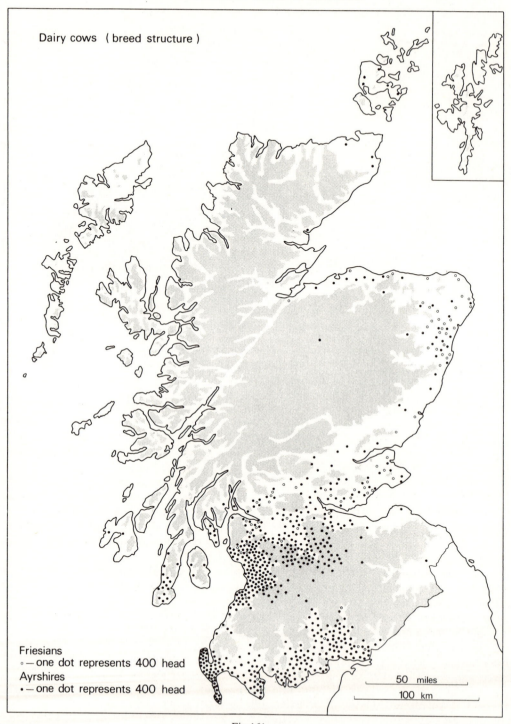

Dairy cows (breed structure)

Friesians
○ — one dot represents 400 head
Ayrshires
• — one dot represents 400 head

50 miles
100 km

Fig 161

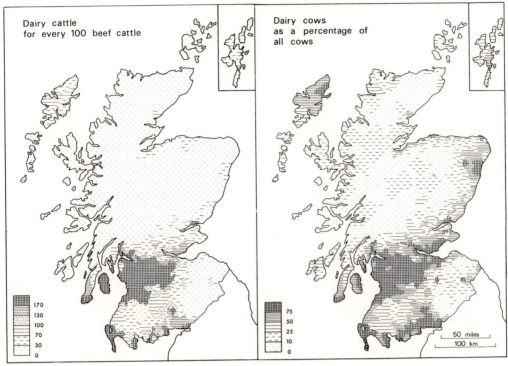

Figs 162–163

DAIRY COWS

Number: 340,389 in 1965; 312,568 in 1972

Dairy cattle per 100 beef cattle: Mean 73·5; SD 118·4; Max 106·9; Min 0·0

Dairy cows as a percentage of all cows: Mean 40·3; SD 31·7; Max 100·0; Min 0·0

Fig 161, showing numbers of dairy cows, is not strictly comparable with other distribution maps in this section since it records, not the number from the agricultural census, but that of cows on milk-selling farms as enumerated in the census conducted by the milk marketing boards, which is about 4 per cent smaller. The reason for using this latter source is that it provides information on the breed structure of the dairy herd, but this distribution does not differ greatly from that of all dairy cows or, indeed, of all dairy cattle.

The distribution of breeds of dairy cows shows another distinctive feature of the Scottish dairy herd, the predominance of the Ayrshire breed, though this is declining; Ayrshires accounted for 78 per cent of cows on farms registered with the milk marketing boards in 1965 and for 58 per cent in 1972. The dominance of Ayrshires was most marked in the main dairying areas, reaching 94 per cent in Wigtownshire in 1965 (though this proportion had fallen to 78 per cent by 1972). Friesians, accounting for 18 per cent of cows, were the second most important breed, outnumbering Ayrshires in several counties, notably in the north-east. Much of the decline in the number of Ayrshires since 1965 has been due to the increasing importance of Ayrshire/Friesian crosses, which accounted for 3 per cent in 1965 and 19 per cent in 1972.

A map of the ratio of dairy to beef cattle also illustrates the importance of the western half of the central lowlands (Fig 162); the relative unimportance of dairying in the eastern lowlands is confirmed, with the exception of west Fife, but dairy cattle were relatively more important in both Lewis and Shetland (although numbers in both areas were small). When dairy cows are plotted as a proportion of all cows, a broadly similar pattern emerges, but it is less clear cut because of the spatial segregation of beef breeding and fattening (Fig 163).

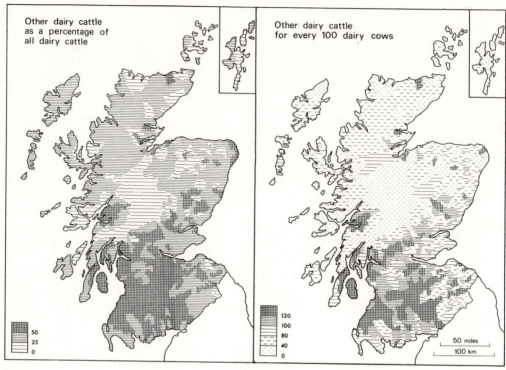

Figs 164–165

OTHER DAIRY CATTLE

Number: 253,209 in 1965; 228,131 in 1972

Other dairy cattle as a percentage of all dairy cattle: Mean 30·2; SD 12·9; Max 83·8; Min 0·0

Other dairy cattle per 100 dairy cows: Mean 65·1; SD 66·3; Max 103·3; Min 0·0

Other dairy cattle are interpreted here as dairy followers other than first-line replacements (ie heifers in calf with first calf). If all herds were self-replacing, the ratio of other dairy cattle to all dairy cattle or to the dairy cows which they will in due course replace would be the same throughout the country, assuming that mortality rates and herd life did not vary. In fact, while 80 per cent of herds in 1969 were self-replacing (or virtually so) the proportion varied considerably throughout the country, for the south-west also provides some replacements for farms elsewhere; in 1969 an estimated 25,000 dairy heifers were supplied to other Scottish farms, mainly from the south-west, especially to farms in north and north-east Scotland which have flying herds, ie those which depend on others for replacements. There are also more local movements within the South West Region, particularly the tendency for occupiers of small farms to summer young stock on hill farms; in 1969 more than a third of all Scottish dairy farms were believed to rent summer grazings and it was estimated that some 13,000 young stock were being reared as stores on non-dairy farms. The relationships between other dairy cattle and other classes of dairy stock are shown in a broad way in Figs 164 and 165. Other dairy cattle accounted for more than 50 per cent of all dairy cattle in most parts of the south-west (Fig. 164), whereas elsewhere in Scotland the proportion was lower, being less than 25 per cent in parts of the Highlands. The high ratio of young cattle to dairy cows in the Southern Uplands (in actual fact, the fringes of the uplands), is the main feature of Fig 165.

MILK SUPPLIES

By comparison with beef cattle, there is comparatively little movement of dairy cattle,

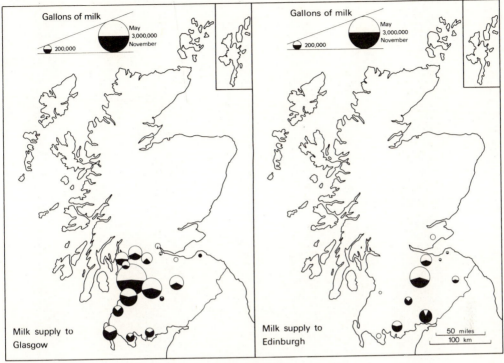

Figs 166–167

whether seasonally or permanently. There is, however, movement of milk to market, and this, as Figs 166 and 167 show, exhibits some seasonal variation. These maps record, for Edinburgh and Glasgow (the main markets), the volume of milk supplied to those cities in May and November 1965. A large part of Glasgow's milk was produced in close proximity to the city in both summer and winter, although a higher proportion of summer milk came from these nearby areas; in winter, more distant producers in south Ayrshire and Galloway provided a larger share (Fig 166). The pattern of Edinburgh's supply was broadly similar (Fig 167), though winter supplies came from Dumfriesshire rather than Galloway.

DAIRY CATTLE AND FARM TYPE

Dairy cattle are also unusual in the extent to which they are concentrated on a single type of farm, dairy farms, and this tendency has become more marked (Table 38). The proportion of other dairy cattle on dairy farms was rather lower. This high degree of concentration on one type may seem at variance with the much higher proportions of holdings with dairy cows, but, as has been noted, many of these animals are merely house cows producing milk for the farm.

TABLE 38
Dairy Cattle and Type of Farming

	Hill sheep	Upland	Rearing with arable	Rearing with intensive livestock	Arable rearing and feeding	Cropping	Dairy	Intensive	Part-time	Scotland
			Percentage of dairy cows on each type of farm							
1965	1	2	2	—	1	2	89	1	2	100
1970	—	2	1	—	—	1	94	—	1	100
			Percentage of other dairy cattle							
1970	—	8	2	—	1	2	83	1	4	100
			Percentage of holdings with dairy cows							
1968	32	24	19	16	21	15	100	4	8	21

Source: *Agricultural Statistics*

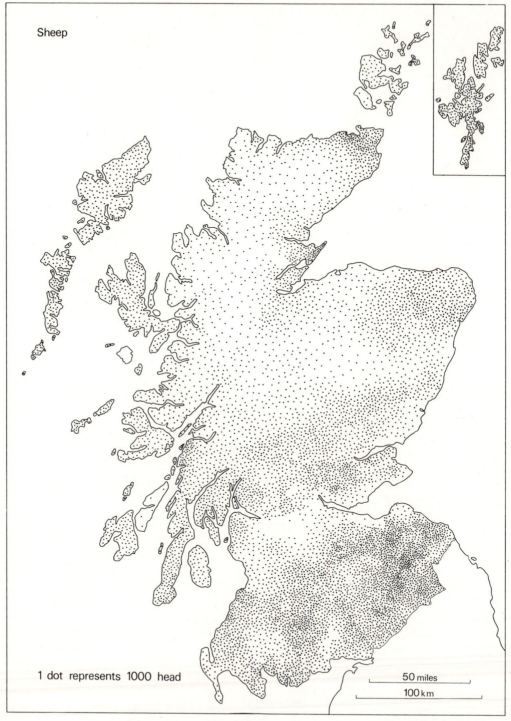

Sheep

1 dot represents 1000 head

50 miles

100 km

Fig 168

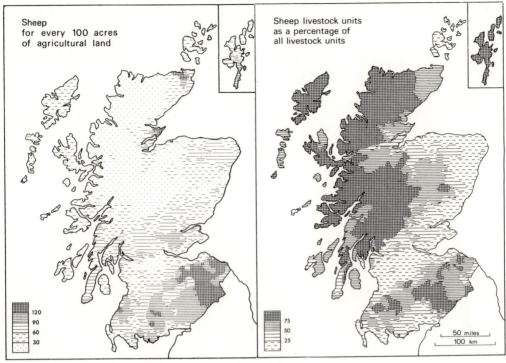

Sheep
for every 100 acres
of agricultural land

Sheep livestock units
as a percentage of
all livestock units

120
90
60
30

75
50
25

50 miles
100 km

Figs 169–170

SHEEP

Number: 8,586,642 in 1965; 7,551,791 in 1972
Sheep per 100 acres of agricultural land: Mean
  66·4; SD 38·4; Max 201·7; Min 0·0
Sheep as percentage of all livestock: Mean 44·8;
  SD 26·8; Max 98·2; Min 0·0

Although sheep (as fat stock, store stock sold to farmers outside Scotland, and wool) contributed only 8 per cent of gross agricultural output in 1965/6 and 11 per cent in 1970/1, they occupy a unique place in Scottish farming; for while they are found on all types of farm, they are most numerous on hill sheep farms, which occupy more than a third of Scottish farmland and where they have exclusive use of a large part of the area under rough grazing. The sheep industry also shows both a distinctive stratification of breeding and fattening and distinctive seasonal differences in distribution.

Sheep are widespread in Scotland and variations in numbers in the uplands reflect the physical suitability of the land for sheep, while differences in the lowlands are more a consequence of competition from other enterprises. In 1965 sheep were most numerous in the

Southern Uplands and in the Tweed basin; they were also found in considerable numbers along the eastern lowlands from Fife to Caithness (Fig 168). They were fewest in the central lowlands and in the Highlands, the former because of competition from dairy cattle, the latter because of the poverty of the upland grazings and the harshness of the environment. Fig 169 highlights the importance of southern (and especially south-eastern) Scotland. However these maps portray a summer distribution; in December, when the sheep population in 1972 was only 71 per cent of that in June, only breeding stock remain on the uplands and the lowlands are relatively more important.

When the relative importance of sheep in summer is measured, the resulting pattern is rather different (Fig 170). While sheep accounted for at least three-quarters of all livestock throughout most of the Highlands, they did so in relatively few places in the Southern Uplands; except in the Tweed valley, the Plain of Caithness and much of Fife, sheep in the lowlands accounted for less than a quarter. For whereas the poorer uplands can be used for no other enterprise, sheep have to compete elsewhere.

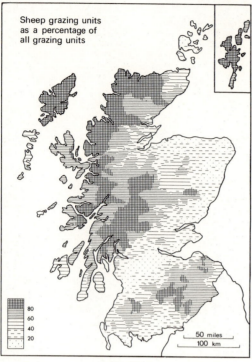

Sheep grazing units as a percentage of all grazing units

80
60
40
20

50 miles
100 km

Fig 171

Sheep as a percentage of all grazing units:
Mean 45·5; SD 28·5; Max 98·4; Min 0·0

Fig 171 provides an alternative assessment of relative importance by showing sheep as a proportion of all grazing units (ie sheep and cattle). This pattern is very similar to that shown in Fig 170, a reflection of the relatively minor part played by pigs and poultry in Scottish farming. In interpreting both this map and other distribution maps of sheep, it is important to bear in mind the large size of most upland parishes (Fig 1) and the fact that many of them include low ground. The predominance of sheep in the uplands would have been shown even more clearly if such parishes included no lowland.

Sheep, whether as lambs or as breeding ewes and their replacements, are widespread on Scottish farms, being found on nearly half the holdings for which returns were made in 1967, ie including all spare- and part-time holdings. They also occurred in large numbers in each Region (Table 39). At this level of generalisation,

their distribution changed little between 1965 and 1972, though numbers fell considerably, and in detail there is evidence of the withdrawal of sheep from the poorer upland grazings.

TABLE 39

Sheep in 1965, 1967 and 1972

	High-land	North East	East Central	South East	South West	Scot-land
Percentage of holdings with sheep in 1967						
	61	40	37	49	39	47
Percentage of sheep in each region						
1965	29	13	14	19	26	100
1972	29	12	14	18	28	100

Source: Agricultural Censuses

As the sheep industry is at present organised, the main controlling factors are the number of sheep which can be supported by the rough grazings in winter and the competitive position of sheep in relation to both beef and dairy cattle in the lowlands. The flocks of breeding sheep on the hills survive on a starvation diet during the winter, the density of stocking being related mainly to the quality of the grazing (though breed is also important); thus, the grass and heather moors of the Southern Uplands carry proportionally many more sheep than the deer sedge and cotton grass moors of the Highlands. To some extent this contrast is exaggerated in Fig 169 because some of the land included on hill sheep farms in the Highlands is either not grazed at all and is agricultural only in name or is grazed only very lightly, and yet has been included in calculations of stocking density; sheep also have to share much of the hill country with grouse and deer, though there is no evidence that these materially affect the number of sheep grazed. These uplands can support more animals in summer when there is abundant growth, and sheep may in fact find it difficult to keep pace with the production of vegetation. Unfortunately, its quality is such that few animals can be fattened on such feed alone. As a result of the low nutritional value of the grazing and of the severity of the environment, losses of breeding stock are higher and the lambing percentage, ie the number of lambs born per 100 ewes, is lower than in the lowlands; and since many of the lambs born on the hills cannot

Table 40

Ewes and Lambs in 1966

	Ewes and gimmers put to the ram in 1965	Lambs bred in 1966	Lambs for breeding	Lambs for feeding	Lambs per 100 ewes	Lambs for feeding per 100 ewes
	In millions					
Hill flocks	2·41	1·78	0·70	1·08	73	45
Upland and lowland flocks	1·44	1·95	0·29	1·66	134	114
All flocks	3·85	3·73	0·99	2·74	97	71

Source: W. J. Carlyle

easily be fattened there, they are sold, mainly in the autumn, for fattening in the lowlands. In addition, female lambs intended as flock replacements may be sent to lowland farms for their first winter (especially to those where cattle are housed in winter and grazing is available for sheep), though the practice appears to be less common than formerly owing to the cost of renting such grazing.

Because of the low lambing percentages on the hills, breeding stock represent a higher proportion of all sheep than in the lowlands (Table 40). Furthermore, because of higher losses and because breeding ewes are drafted out of hill flocks after three or four breeding seasons, more than half of them to produce one or two more crops of lambs in the lowlands, flock replacements also represent a higher proportion of all sheep in the uplands. This relationship is recorded in a very broad way in Table 40, which shows the cumulative effect of lower lambing percentages and larger flock replacements on the number of lambs available for feeding.

Sheep are also bred in the lowlands and such flocks depend largely on the uplands for a supply both of cast ewes (which form the basis of the flying (ie non-permanent) flocks, to be sold in spring with lambs at foot) and of ewe lambs which are surplus to the needs of the upland flocks. The general pattern is one in which ewes are crossed with rams of lowland breeds, notably the Border Leicester and the Down breeds (especially the Suffolk and the Oxford Down), to produce lambs of a kind which meets the needs of the market and exploits the more favourable conditions of the lowlands. First crosses, usually with a Border Leicester ram, are

the Greyface (from a Blackface ewe) and the Half-Bred (from a Cheviot ewe); the progeny of these are in turn crossed with Down rams to produce Down Crosses, though few female lambs of this cross are retained for breeding (Table 42). Distributions and sequences of movement vary with the different breeds and are discussed in relation to Figs 172 and 173, which show the distribution of the two main breeds, the Blackface and the Cheviot, but this general stratification, marked by movement from hill to lowland of ewes and ewe lambs for breeding and of wether and, to a lesser extent, ewe lambs for fattening, is the most distinctive feature of sheep farming in Scotland. In general, the lower the elevation and the lower the proportion of rough grazing, the higher the lambing percentage and the higher the proportion of lambs which are fattened rather than sold as stores; thus, whereas lambing rates on the hills are between about 65 and 100 lambs per 100 ewes, these values range from between about 110 and 145 in the uplands, and generally exceed 145 in the lowlands. Such differences are in part due to the breeds used, but they are also due to environmental differences and the same breed will have different lambing rates under different conditions; thus, while Blackface ewes had lambing rates of only 65 per cent in Inverness and further north, they achieved rates of 100 per cent in Aberdeen, Banff and Kincardineshire. In a very broad sense these differences are reflected in Table 41, which shows the proportion of breeding ewes and of lambs in each region. Thus, the Highlands had a lower proportion of the lambs in Scotland than of the ewes, whereas the other regions had more.

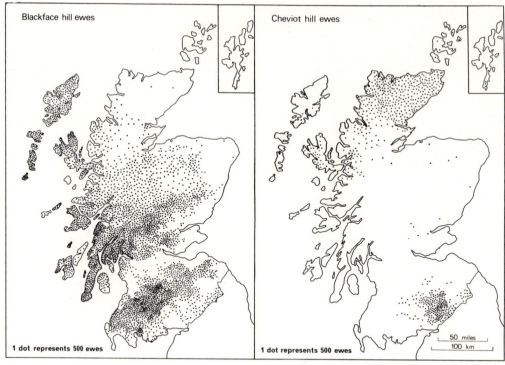

Figs 172–173

TABLE 41

Ewes and Lambs in 1965 and 1972

	High-land	North East	East Central	South East	South West	Scot-land
Percentage of ewes in each region						
1965	31	12	14	17	26	100
1972	31	10	14	17	28	100
Percentage of lambs in each region						
1965	23	15	14	21	27	100
1972	24	13	14	20	29	100

Source: Agricultural Censuses

BREEDS OF SHEEP

Details of the breeds of sheep are available only for those eligible for Hill Sheep Subsidy, and Figs 172 and 173 show the distribution of the two principal breeds in 1967. Blackface ewes were by far the most numerous, accounting for 78 per cent of those eligible for subsidy. They were widely distributed throughout the uplands but were relatively more important on the poorer grazings of the west; they were virtually absent from Caithness and Sutherland and from Orkney and Shetland (where nearly all ewes were of the Zetland breed which accounted for a further 4 per cent). Different types of Blackface are recognised, but while these play some part in determining the detailed geography of sheep farming in Scotland, they will not be considered here. Blackface ewes are also the source of Greyface lambs, which are produced mainly on upland farms where there is a regular breeding flock of Blackface ewes and a Border Leicester ram.

Cheviot ewes accounted for 16 per cent of eligible ewes and were to be found in two main areas in 1967, the headwaters of the Tweed and Teviot, and Caithness and Sutherland (Fig 173). In fact, as this distribution suggests, there are two different types which have almost reached the stage of separate breeds, the North Country Cheviot, a larger animal found chiefly in Caithness, Orkney and the eastern parts of

Sutherland and Ross and Cromarty, and the South Country Cheviot, found mainly on the grass moors of the eastern Borders. Large numbers of Cheviot ewe lambs are sold to farmers in other parts of Scotland and in England to provide replacements for flocks producing Half-Bred lambs; in fact, nearly as many Cheviot ewe lambs as Blackface lambs are marketed, though Blackface ewes outnumber Cheviot ewes by five to one. Most of these Cheviot ewe lambs come from North Country Cheviot flocks. Many more Half-Bred ewes are also marketed than either Blackface or Greyface, despite the larger numbers of the latter breeds. Large numbers of both ewes and ewe lambs are sold to English farmers, though the Half-Bred is losing ground in England to other crosses which are smaller and have lower feed requirements.

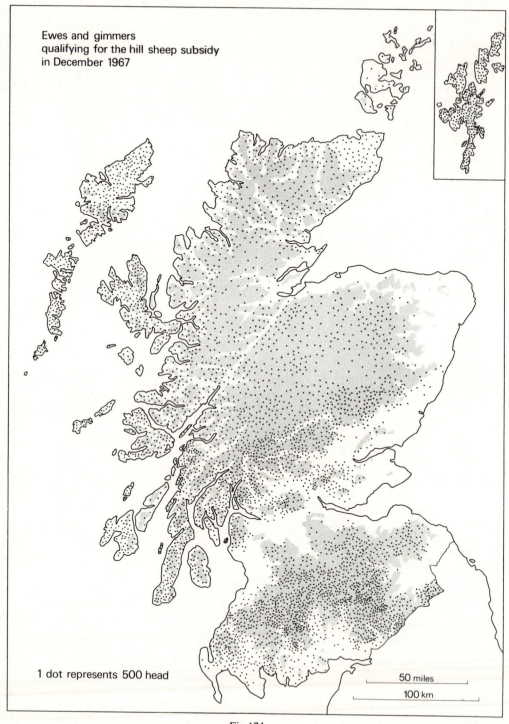

Ewes and gimmers
qualifying for the hill sheep subsidy
in December 1967

1 dot represents 500 head

50 miles

100 km

Fig 174

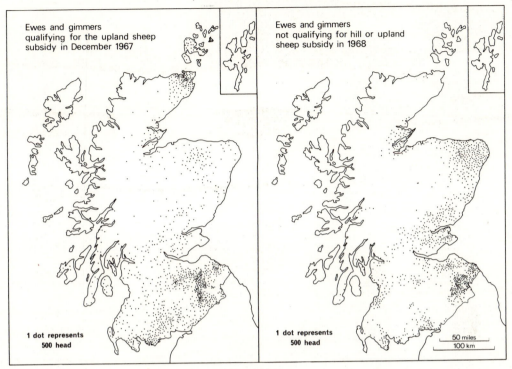

Figs 175–176

BREEDING EWES

One indication of this stratification of sheep farming in Scotland is provided by the distribution of ewes and gimmers attracting full- and half-rate Hill Sheep Subsidy and those on which no subsidy is payable. Full-rate subsidy is payable in respect of flocks of hardy hill breeds (Blackface, Cheviot and Zetland) maintained throughout the year on rough grazings. Most are in pure-bred flocks and Fig 174 shows their widespread distribution throughout the uplands and the islands. Such ewes numbered some 2,448,000 in 1967, or about 68 per cent of all ewes and gimmers.

Ewes and gimmers qualifying for the half-rate subsidy, introduced in 1968, numbered some 542,000, or about 14 per cent of the total breeding flock. Such ewes were to be found mainly in cross-bred flocks throughout the upland margins, on farms with a higher proportion of improved land and capable of both providing some winter feed and fattening a proportion of the lambs reared. These sheep were particularly prominent in 1967 on the higher land of the Tweed, in Dumfriesshire and in the plain of Caithness (Fig 175).

The third category, those in flocks not eligible for either full or half-rate subsidy, have been identified by assuming that subsidy was claimed on all those animals which were eligible; such unsubsidised sheep numbered some 680,000 or about 18 per cent of the breeding flock. These were mainly in cross-bred flocks and were found throughout the lowlands; but they were most numerous in the Merse and in north-east Scotland (Fig 176).

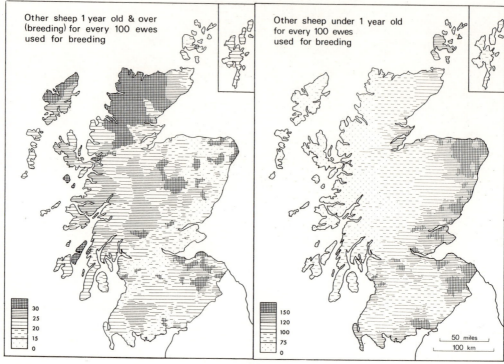

Figs 177–178

Flock Replacements

Other sheep (1+) for breeding per 100 ewes: Mean 20·9; SD 12·6; Max 151·8; Min 0·0

As with many maps which record ratios between different classes of livestock, that of the ratio of other sheep one-year old and over intended for breeding to all breeding ewes shows a patchy distribution (Fig 177). In general, proportions were highest in the uplands where mortality is highest (though the respective death rates were about 6 and 3 per cent) and where ewes are drafted out after three or four breeding seasons for further breeding elsewhere or for slaughter. Ratios were generally lowest in the lowlands, though there are some anomalous high values. Table 42 provides complementary information and shows how the proportion of lambs for breeding varied both regionally and by breeds; values were generally highest in the Highlands, and the three hill breeds all show high proportions. The Half-Bred, which provides breeding stock for Scottish as well as English farms, had a higher proportion of replacements than the Greyface, while flocks producing Down Cross lambs, the final stage in the stratification of Scottish sheep farming, bred very few lambs for replacements.

Lambing Percentages

Lambs per 100 breeding ewes: Mean 118·1; SD 29·9; Max 189·9; Min 0·0

The lambing percentage cannot strictly be calculated from the number of ewes in June owing to losses during the winter; furthermore, while few lambs have been sold by this date, except those sold with their mothers in flying flocks, not all lambs have yet been counted. Nevertheless, the relationship does give a broad indication of lambing percentages and shows the contrast between the lowlands, where there were often 150 or more lambs per 100 ewes, and the uplands where there were generally less than 100, especially in the western Highlands and in the islands (Fig 178).

TABLE 42

Breeding Ewes and Flock Replacements in 1966

	Highland	North East	East Central	South East	South West	Scotland
		In thousands				
Breeding ewes	1,137	418	490	627	932	3,604
Other sheep 1+ for breeding	287	85	105	151	327	983
Lambs retained for breeding	368	115	111	152	237	983
		Percentage of lambs for breeding				
All breeds	47	23	20	19	22	26
Blackface	58	49	45	44	36	46
Cheviot and Zetland	48	40	25	31	36	41
Greyface	19	11	6	6	6	7
Half-Bred	33	44	—	35	25	36
Down Crosses	—	3	3	5	2	3

Sources: W. J. Carlyle and *Agricultural Statistics*

FATTENING

The information provided by the June census on sheep, unlike that on cattle, throws relatively little light on fattening, since this largely takes place within the same census year; no mappable data are therefore available on the location of the different stages of rearing and fattening. Nevertheless, there are considerable movements of sheep, both locally, from hill and (to a lesser extent) upland farms to low ground farms, and regionally, from one part of Scotland to another and from Scotland to England. Table 43 shows the differences by regions and by breeds in the numbers of lambs bred for feeding, though it should be noted that these regions are not very satisfactory for analysing spatial distributions and have been retained only because of the valuable information which has been assembled by them; the North East Region, which includes not only the counties from Kincardine to Nairn, but also Caithness and Orkney, is particularly unsatisfactory for the analysis of movements of

sheep. It should also be noted that the figures in Tables 43–5 are based on market sales, and it is believed that perhaps a tenth of all lambs are sold privately or exchanged between linked farms. In all, some 542,000 of these lambs were transferred from the region in which they were born to be fattened elsewhere, while 270,000 lambs were brought from other regions for fattening, leaving a net loss of 266,000 lambs to English farms. Table 44 shows the numbers leaving and entering each region, by breeds in 1966, and confirms the roles of the Highlands as a reservoir of lambs for feeding and of the East Central Region as a receiving area; such losses include not only those to other regions but to England. Other regions, though net exporters on balance, were net importers of lambs of some breeds, and an analysis of movements within Scotland would show that the South East Region was a net importer from other regions.

As Table 45 shows, east and south Scotland were the main areas for fattening. In all regions

TABLE 43

Lambs Bred for Feeding in 1966

Breed	Highland	North East	East Central	South East	South West	Scotland
		In thousands				
All breeds	407	396	432	630	864	2,729
Blackface	195	21	118	90	313	737
Cheviot and Zetland	108	44	3	46	45	246
Greyface	21	84	163	72	228	568
Half-Bred	2	65	3	68	45	183
Down Crosses	81	182	145	354	233	995

Source: W. J. Carlyle

TABLE 44

Inter-regional Transfers of Lambs for Feeding in 1966

	Highland	North East	East Central	South East	South West	Scotland
			In thousands			
			All breeds			
Leaving	164	77	2	170	129	542
Entering	—	33	103	84	56	270
Net gain (+) or loss (−)	−164	−44	+101	−86	−73	−266
Lambs fed in region	243	352	533	544	791	2,463
			Blackface			
Leaving	76	2	1	14	68	161
Entering	—	5	29	30	38	102
Net gain (+) or loss (−)	−76	+3	+28	+16	−30	−59
Lambs fed in region	119	24	146	106	283	678
			Cheviot and Zetland			
Leaving	69	22	—	17	4	112
Entering	—	23	15	16	10	64
Net gain (+) or loss (−)	−69	+1	+15	−1	+6	−48
Lambs fed in region	39	45	18	45	51	198
			Greyface			
Leaving	10	24	1	12	43	90
Entering	—	2	38	23	6	68
Net gain (+) or loss (−)	−10	−22	+37	+11	−37	−22
Lambs fed in region	11	62	200	83	191	546
			Half-Bred			
Leaving	1	17	1	25	5	48
Entering	—	—	4	10	—	14
Net gain (+) or loss (−)	−1	−17	+3	−15	−5	−34
Lambs fed in region	1	48	7	53	40	147
			Down Cross			
Leaving	8	12	—	102	9	131
Entering	—	3	17	5	2	27
Net gain (+) or loss (−)	−8	−9	+17	−97	−7	−104
Lambs fed in region	73	173	162	257	226	891

Source: W. J. Carlyle

most lambs were fattened on the farms on which they were born and movements from other regions accounted for a maximum of 19 per cent of those fattened (Eastern Region); there were, however, considerable differences between breeds. Thus the Cheviot and Zetland breeds (not separately considered) had the lowest proportion of home-fattened lambs and Down Crosses the highest proportion. These differences are only in part a reflection of the stratum occupied by a particular breed, for rates of maturing are also important; for example, Blackface lambs can be fattened early on rape, an upland crop (Fig 82), while Cheviot lambs mature more slowly and are fed on turnips and other arable crops. It should, however, be noted that the figures for some breeds in some regions are small and should be interpreted with caution (Table 44). It is similarly understandable that the different breeds should vary in respect of the importance of imported stores for feeding; lambs from Cheviot flocks were most likely and

TABLE 45

Lambs Fed in Each Region in 1966

	Highland	North East	East Central	South East	South West	Scotland
Lambs for fattening (all breeds) in thousands						
	243	352	533	544	791	2,463
Percentage fed on farms on which they were born						
All breeds	76	58	52	65	50	57
Blackface	78	42	37	54	34	46
Cheviot and Zetland	39	18	6	40	29	29
Greyface	45	21	45	53	42	42
Half-Bred	100	71	14	53	28	51
Down Crosses	96	80	77	80	85	82
Percentage of lambs for fattening imported from other regions						
All breeds	—	9	19	15	7	11
Blackface	—	21	20	28	13	15
Cheviot and Zetland	—	51	83	36	20	32
Greyface	—	3	19	28	3	12
Half-Bred	—	—	57	19	—	10
Down Crosses	—	2	10	2	1	3

Source: W. J. Carlyle

those from Down Cross flocks least likely to be moved to farms in other regions for fattening. Cheviot and Half-Bred lambs move from north Scotland to East Central and South East Scotland; movements of Greyface lambs, on the other hand, are more local.

Systems of feeding varied considerably throughout Scotland, reflecting mainly the breed of sheep being fattened, the time of year at which they were fattened and the availability of feed. Table 46 records the evidence of a sample survey in 1965/6. It shows that all systems were practised in each region, but that a majority of sheep was fattened on grass alone in the Highlands and in South West Scotland. In no region was a majority of lambs fattened by folding on crops, though folding was relatively more important in East Central and South East Scotland, the principal areas for arable farming.

There were also considerable variations in seasons and in breeds. Thus, while only 14 per cent were folded on roots over the whole year, 44 per cent were fattened in this way between December and February (a figure which reached 90 per cent in the South East Region) and 34 per cent between March and May. As Table 47

shows, home-reared lambs tended to be fattened earlier than those purchased from other farms. The table also confirms the assertion that Blackface lambs mature earlier than Cheviot lambs; for 76 per cent of Blackface lambs were fattened between August and November, compared with only 11 per cent of Cheviot lambs; Greyface and Half-Bred lambs tended to follow the same patterns as their respective parents, though the contrast between them was less marked.

SHEEP AND FARM TYPE

Almost two-thirds of the breeding flock were found on hill sheep and upland farms in 1970, although such sheep occupied a much more important place relatively on hill sheep farms; a further 13 per cent were on spare- and part-time holdings. Owing to higher lambing percentages and lower losses in the lowlands, the share of sheep and lambs under one-year old was considerably lower on hill farms and that of lowland types somewhat greater. The pattern in 1965 was broadly similar, though information was available only for the category 'other sheep', which include both flock replacements and older

TABLE 46

Systems of Fattening in 1965/6

System	Highland	North East	East Central	South East	South West	Scotland
	Percentage of lambs fattened in each region					
Grass only	65	18	25	38	68	42
Grass with other (mainly turnips)	29	79	30	29	14	33
Folded roots	—	2	22	29	6	14
Folded green fodder (mainly rape)	3	—	23	3	11	10
Fed in yards or sheds	3	1	—	1	1	1
All systems	100	100	100	100	100	100

Source: *Scottish Agricultural Economics*, Vol 17

TABLE 47

Seasonality of Fattening in 1965/6

	Home-bred and fattened	Purchased for fattening	All breeds	Blackface	Greyface	Cheviot and Zetland	Half-Bred	Down Crosses
	Percentage of lambs fattened							
April–July	12	—	8	—	2	3	—	15
August–September	29	6	21	22	20	6	19	25
October–November	31	31	31	54	29	5	20	28
December–February	17	30	21	8	27	16	44	20
March–August	11	33	18	16	22	70	17	12

Source: *Scottish Agricultural Economics*, Vol 17

TABLE 48

Sheep and Farm Type in 1965, 1968 and 1970

	Hill sheep	Upland	Rearing with arable	Rearing with intensive livestock	Arable rearing and feeding	Cropping	Dairy	Intensive	Part- and spare-time	All
	Percentage of breeding ewes on farms of each type									
1965	31	23	14	1	3	6	10	—	12	100
1970	36	31	8	1	2	4	6	—	13	100
	Percentage of other sheep on farms of each type									
1965	23	22	18	1	5	8	12	—	11	100
	Percentage of sheep and lambs under one-year old									
1970	28	34	11	1	3	6	7	—	10	100
	Percentage of holdings with breeding ewes									
1968	100	79	62	42	50	36	35	7	37	43

Source: *Agricultural Statistics*

sheep as well as lambs under one year old. There has also been some modification of farm classification between these dates, which helps to account for the different proportions of upland and rearing with arable farms (see p 172).

In similar fashion, the proportion of holdings with breeding ewes also varies markedly with type of farm, though, except for intensive farms, proportions are surprisingly high, a reflection of the widespread distribution of sheep in Scotland.

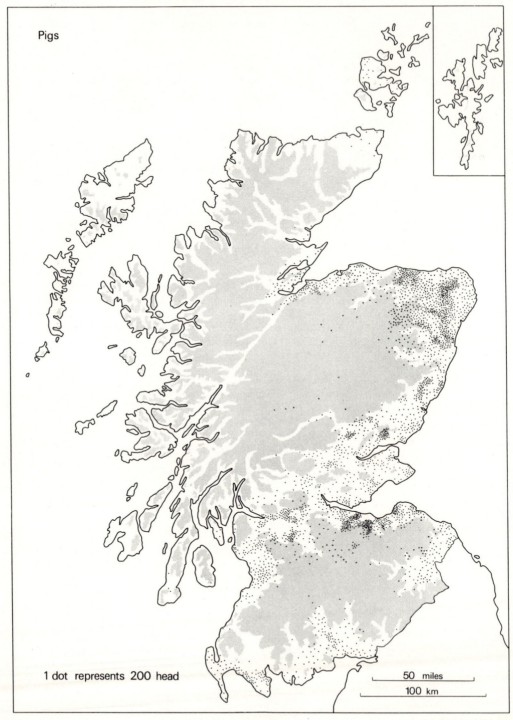

Pigs

1 dot represents 200 head

50 miles

100 km

Fig 179

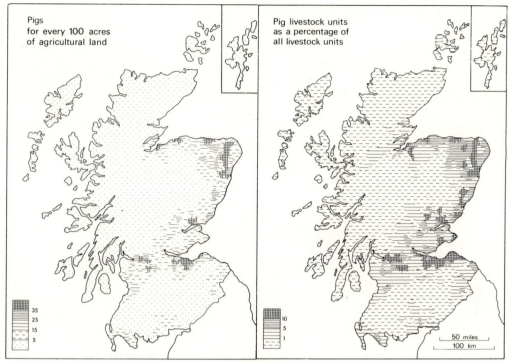

Figs 180–181

PIGS

Number: 553,321 in 1965; 661,833 in 1972

Pigs per 100 acres of agricultural land: Mean 7·4; SD 13·4; Max 155; Min 0·0

Pigs as a percentage of all livestock units: Mean 3·2; SD 5·2; Max 58·5; Min 0·0

Pigs accounted for 7 per cent of agricultural output in both 1965 and 1970, and were found on only 12 per cent of holdings in 1967, a ratio which was lowest in the Highlands and highest in the North East. Their distribution in 1965 was very patchy, with concentrations around the major cities (especially Edinburgh) and in the north-east, and with smaller numbers elsewhere throughout the lowlands of the east and south (Fig 179). This patchy distribution is confirmed by Fig 180 which shows the same important areas as having 25 or more pigs per 100 acres. In relative terms, too, the urban fringes were the most important areas, accounting for 10 per cent or more of all livestock (Fig 181); nearly everywhere in the north-east the proportion

exceeded 5 per cent. The North East was, in fact, the principal region and its importance has become more marked in recent years.

Pigs were mostly found on four types of farm: intensive (with 33 per cent of breeding sows and 42 per cent of other pigs two-months old and over in 1970); rearing with intensive livestock (with 20 per cent and 17 per cent); cropping (with 19 per cent and 17 per cent); and dairy farms (with 10 per cent and 12 per cent). They were also found in smaller numbers on farms of all types.

TABLE 49

Pigs in 1965, 1967 and 1972

	High-land	North East	East Central	South East	South West	Scot-land
	Percentage of holdings with pigs					
1967	1	21	16	18	12	12
	Percentage of pigs in each region					
1965	3	41	18	17	22	100
1972	3	51	17	13	16	100

Source: Agricultural Censuses

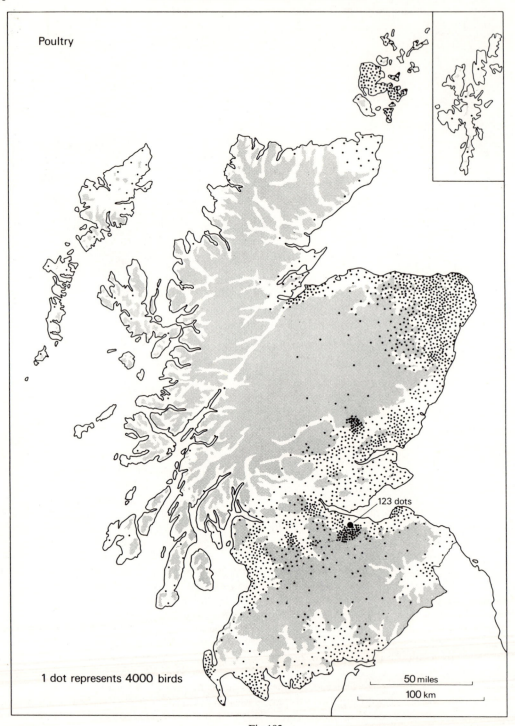

Poultry

123 dots

1 dot represents 4000 birds

50 miles

100 km

Fig 182

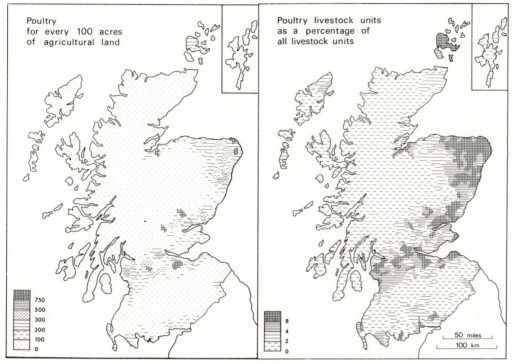

Figs 183–184

POULTRY

Number: 9,116,945 in 1965; 14,108,329 in 1970
Poultry per 100 acres of agricultural land: Mean
    121; SD 377; Max 7,228; Min 0
Percentage of all livestock units: Mean 4·5; SD
    5·8; Max 53·5; Min 0·1

Poultry as enumerated in the agricultural census
include fowl, ducks and turkeys, which accounted
for 98, 1·6 and 0·5 per cent respectively of all
poultry in 1965. In that year they provided 2·8
per cent of gross agricultural output, while eggs
produced a further 6·2 per cent (compared with
3·9 and 6·0 per cent respectively in 1970).
Poultry occur on a large proportion of holdings
(58 per cent in 1965 and 51 per cent in 1970), but
many of these flocks were not commercial and
were kept for domestic purposes; an increasing
proportion was to be found on a small number
of very large units (as measured by the number
of birds), so that their distribution was patchy
(Fig 182). Poultry, now kept largely in artificial
environments, were to be found chiefly in the
lowlands, especially in the north-east (with 23

per cent in Aberdeenshire alone) and the central
lowlands (10 per cent in Midlothian), and also
in the south-west (Table 50). Orkney, where
poultry keeping developed during the period
between the two world wars following the break-
up of estates, was unique among the crofting
counties in having large numbers of poultry.

Fig 183 confirms this patchy distribution and
the importance of the north-east and the central
lowlands, while Fig 184 shows that, relatively,
poultry were more important in the north-east,
with 8 per cent or more of all livestock units. On
both maps Orkney stands out as anomalous
among the islands.

TABLE 50

Poultry in 1965 and 1972

	High-land	North East	East Central	South East	South West	Scot-land
	Percentage in each region					
1965	4	37	19	15	25	100
1972	2	17	43	20	18	100

Source: Agricultural Censuses

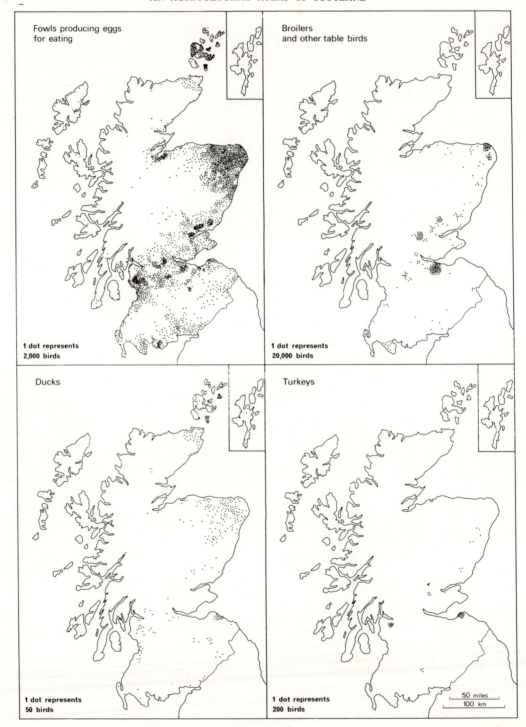

Fowls producing eggs
for eating

1 dot represents
2,000 birds

Broilers
and other table birds

1 dot represents
20,000 birds

Ducks

1 dot represents
50 birds

Turkeys

1 dot represents
200 birds

50 miles
100 km

Figs 185–188

FOWLS FOR PRODUCING EGGS:

Number: 5,587,363 in 1965; 6,633,359 in 1972

Fowls kept primarily for producing eggs accounted for 61 per cent of all poultry in 1965 and 47 per cent in 1972; their distribution is thus fairly similar to that of poultry as a whole (Fig 185). A majority of all holdings kept some fowls for egg laying, but many of these were non-commercial enterprises; in 1965 some 46 per cent of laying flocks had fewer than twenty-five birds, and the 1970 figure was 50 per cent, despite a very considerable fall in the number of flocks through the elimination of many small holdings from the census and a long-established trend towards the reduction in the number of units. At the other extreme, 65 units with 5,000 or more birds had 16 per cent of the total flock in 1965 and there were 141 such units in 1970, accounting for 62 per cent of all birds. The distribution is thus even more patchy than the map suggests, for a large proportion of the flock in a given parish may be concentrated on a single holding occupying only a few acres; additionally, it is a weakness of dot maps that a large number of dots cannot be located wholly within the boundaries of the parish to which they refer. The distribution in the north-east is more uniform, in that only 12 per cent of hens and pullets were in such flocks in 1965, compared with 26 per cent in the East Central Region (Table 51). Laying flocks were associated with a wide variety of farm types in 1965; but, whereas only 14 per cent were with intensive types in 1965, this proportion had reached 58 per cent by 1970.

TABLE 51

Fowls and Turkeys in 1965 and 1972

	High-land	North East	East Central	South East	South West	Scotland
Percentage of hens and pullets in laying flocks in each region						
1965	6	44	14	8	28	100
1972	4	23	38	9	26	100
Percentage of broilers in each region						
1965	1	25	28	27	19	100
1972	—	13	49	31	6	100
Percentage of turkeys in each region						
1965	2	23	16	26	33	100
1972	—	14	4	17	64	100

Source: Agricultural Censuses

BROILERS AND OTHER TABLE BIRDS

Number: 2,792,240 in 1965; 6,234,596 in 1972

The distribution of broilers (which account for the great majority of birds in this class) and other table birds was even more patchy, for such birds were recorded on only 3 per cent of agricultural holdings in both 1965 and 1970; they were largely confined to the central lowlands and the north-east (Fig 186). Moreover, the concentration of birds on a few holdings was much more marked, ten holdings with 50,000 or more birds accounting for 53 per cent of all broilers in 1965 and twenty-three accounting for 78 per cent in 1970. Furthermore, the throughput of broilers is considerably larger than totals for such birds on 4 June suggest, for the average age of slaughter of broilers is about sixty-five days and 11,821,000 were killed in 1964/5. The keeping of broilers was strongly associated with intensive farms, which accounted for 45 per cent of all broilers in 1965 and 73 per cent in 1970.

DUCKS

Number: 45,027 in 1965

Ducks were of minor and declining importance in 1965, accounting for 0·5 per cent of poultry, and ceased to be enumerated in the June census of 1968. They were strongly associated with Aberdeenshire and Dumfriesshire, each of which had 20 per cent of all ducks in 1965 (Fig 187).

TURKEYS

Number: 141,998 in 1965; 202,615 in 1972

Numbers of turkeys, by contrast, have been increasing slowly, though accounting for only 1·6 per cent of poultry in 1965 and 1·4 per cent in 1972. Turkeys were both highly concentrated on a few holdings, being recorded on only 1 per cent of all holdings, and highly localised, with 15 per cent in Aberdeenshire, 24 per cent in East Lothian and 19 per cent in Renfrewshire (or 11 per cent, 17 per cent and 62 per cent respectively in 1972). Turkeys were strongly associated with intensive farms which accounted for 73 per cent of all the birds recorded in 1970.

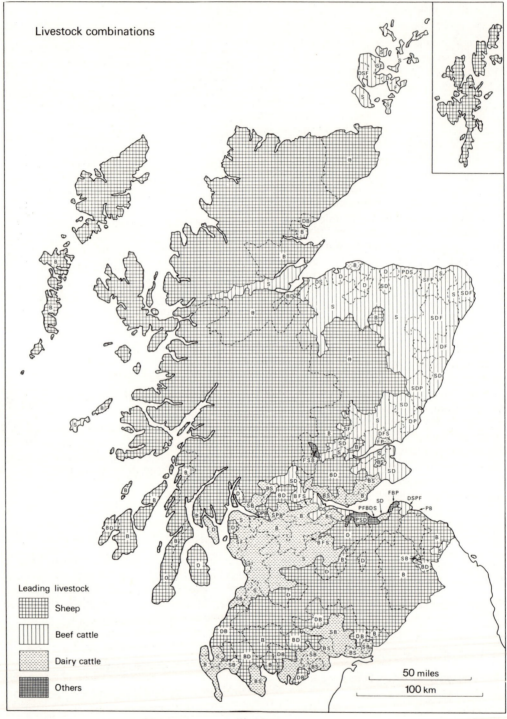

Livestock combinations

Leading livestock

- Sheep
- Beef cattle
- Dairy cattle
- Others

50 miles
100 km

Fig 189

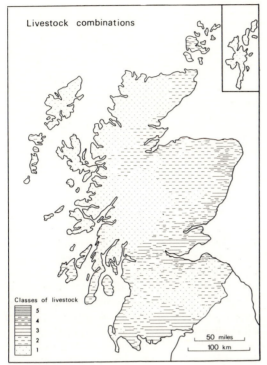

Fig 190

LIVESTOCK COMBINATIONS

As with the map of crop combinations (Fig 106), Fig 189 is an attempt to synthesise the preceding information in this section on the regional distribution of the various kinds of livestock; the same method is used of identifying the closest approach to ideal combinations. The shadings show the first-ranking (leading) class of livestock and the letters indicate, in descending rank order, the other livestock in the combinations identified for each parish; other classes of livestock may,

of course, be present, but not in sufficient numbers to be included in the livestock combination. Five classes of livestock are identified by these letters: B, beef cattle; D, dairy cattle; F, poultry; P, pigs; and S, sheep. Throughout the uplands, sheep were the leading livestock, and over the greater part of the western Highlands and the higher reaches of the Southern Uplands they were the only class of livestock to appear in livestock combinations. Along the eastern margins of the Highlands, beef cattle were usually the second-ranking livestock, but in many parishes in the south-west, their place was taken by dairy cattle. In the lowlands, beef cattle were the leading type between the Forth and the Moray Firth, whereas dairy cattle were the first-ranking livestock over most of the central lowlands and in the south-west. Orkney was again anomalous among the major island groups, with beef cattle rather than sheep as the dominant livestock. The Merse, too, differed from the other major areas of crop production in that sheep were the leading livestock nearly everywhere. Pigs and poultry were the leading livestock in only a few parishes, mainly in the Lothians.

It is difficult to grasp the complexity of Fig 189, although the shadings used to identify the leading class of livestock present a clear, simple picture, and so further clarification is given by Fig 190, which shows the number of classes of livestock recorded in these combinations. Combinations comprising only the dominant class of livestock were largely confined to the Highlands and Southern Uplands and to the great majority of the islands (except Orkney). The most complex combinations occurred in the north-east, the Lothians and the lowlands of Perth and Angus, while the margins of the uplands generally had two-class combinations.

5

Agricultural Enterprises

Although the competitive power of the different kinds of stock and crops has been assessed in the previous section the assessments are of only limited value—made by computing the proportion of the particular class of livestock of which it is a subset, eg other beef cattle under one-year old as a percentage of all beef cattle or breeding ewes as a percentage of all sheep, the latter by calculating the percentage of the land under tillage or that under groups of related crops, such as cereals. However, these assessments are more important as a contribution to the whole range of agricultural activities on each holding or in each area. As has already been indicated, it is impossible, in any meaningful sense, to compare directly acreages of crops and numbers of livestock, which must therefore be converted to some common base. For present purposes, the base used is that of standard labour requirements, chiefly because they have been adopted for official use by the Department of Agriculture and Fisheries for Scotland and because such factors change less rapidly than, say, prices. Each item in the June census relating to the use of land or to livestock is multiplied by an appropriate factor to give the number of standard man-days for that item; these are then grouped to form major enterprises, such as dairying, though for some purposes even a single crop, such as wheat, can be regarded as an enterprise.

These factors are not, however, fixed for all time; as agricultural productivity improves, through greater mechanisation and in other ways, so the labour requirements for individual crops and classes of livestock decline. It is therefore necessary to revise the standard labour requirements from time to time and this has been done at intervals of approximately three years. Such revisions make for a more realistic assessment of the different enterprises, but they have one major disadvantage, in that they make it impossible to compare absolute values over a period of years; relative values, on the other hand, should not be as greatly affected unless the rates of change have differed widely among the the individual factors. The resulting man-days can then be added to give the total number of standard man-days (smd) for each holding, parish or larger administrative unit and, when calculated in relation to the total area of farmland, give a measure of the intensiveness of agricultural activity (Fig 191). The proportions of this total which can be attributed to each enterprise can also be calculated to give a measure of the relative importance of that enterprise throughout the country (Figs 192–201).

It must, of course, be appreciated that these are standard factors and do not represent the actual labour requirements on each holding, which will vary with the topography, the size of fields and the layout of farms; they are also likely to vary with the personal attributes of the farmer and of any employees, particularly their health, age and physical condition, and with the machinery at the farmer's disposal. It is likely, therefore, that the actual labour used will be greater than these standard factors where the terrain is steep or broken, where fields are small and narrow and where farms are small and fragmented, or where the average age of farmers is high and they depend primarily on their own labour, as in many areas of declining population in the hills. Such discrepancies between standard and actual are particularly likely in the growing and harvesting of crops, which will require considerably more labour (proportionately) on a small croft than on a large arable farm. Fortunately, the importance of such differences is lessened by the fact that only small acreages tend to be grown where conditions are unsuitable for crop growth and labour requirements are high; for, as Fig 51 shows, the growing of crops is highly concentrated in those parts of the country which are best suited to crop production. Labour requirements for livestock are likely to vary less widely than those for crops and, in any case, these enterprises are also highly regionalised. The procedure thus provides a useful basis for comparing the different enterprises and some indication of absolute variations in agricultural intensity.

As Table 52 shows, there have been quite considerable revisions in the standard man-days used in Scotland, even over as short a period as six years, during which the total number of smds calculated on the basis of standard factors fell by about 30 per cent. For most field crops, they have fallen by less than 40 per cent, but for some crops, such as kale and cabbage for stock-feeding, the decline has been much greater, while in a few, such as beans for stockfeeding, standard factors have been increased. The changes among horticultural crops have been greater; thus, the factors used for broad beans and for peas for canning, freezing and drying in 1968 were only 16 per cent of those used in 1962, while for rhubarb and leeks the corresponding figure was 267 per cent. Livestock man-days seem to have declined rather less, though these changes, too, range from laying fowl, for which standard labour requirements were only 32 per cent of those in 1962, to other fowl, where the corresponding figure was 200 per cent. Of course, these elements are not of equal importance, for some items make only a minor contribution to agricultural production; but caution must be exercised in interpreting changes over time.

Table 52 *on page* 158

TABLE 52

Standard Man-days in 1962, 1965 and 1968

	1962	1965	1968	1968 as a
Crops		smd per acre		percentage of 1962
Wheat, barley	3·125	2·5	2·0	64
Oats, mixed grain (threshing)	3·75	3·5	3·0	80
Potatoes, seed	18·75	19·0	17·0	91
Potatoes, ware	18·75	19·0	15·0	80
Turnips and Swedes (for stock)	12·5	11·0	9·0	72
Mangolds	18·75	11·0	11·0	59
Sugar beet	18·75	12·5	10·0	53
Kale, cabbage (for stock)	6·25	6·0	3·0	48
Rape	1·875	1·5	3·0	53
Beans (for stock)	1·875	1·5	1·0	160
Mashlum, rye, vetches etc	4·375	3·0	3·0	69
Other crops	6·25	3·0	3·0	48
Bare fallow	0·25	0·5	0·5	200
Grass				
Mowing grass under 7 years	2·75	1·5	1·25	45
Mowing grass, 7 years and over	2·5	1·5	1·0	40
Other grass, under 7 years	0·4	0·25	0·25	63
Other grass, 7 years and over	0·25	0·25	0·25	100
Horticulture				
Peas for canning, freezing etc	18·75	4·5	3·0	16
Broad beans	18·75	30·0	3·0	16
Leeks	18·75	30·0	50·0	267
Turnips and swedes	25·0	21·0	20·0	80
Cabbage and savoys	25·0	17·0	20·0	80
Brussels sprouts	25·0	25·0	20·0	80
Cauliflower, broccoli	25·0	17·0	20·0	80
Beetroot	31·25	35·0	30·0	96
Lettuce	31·25	10·0	30·0	96
Carrots	31·25	17·0	12·0	38
Rhubarb	18·75	50·0	50·0	267
Other and mixed vegetables	37·5	50·0	50·0	133
Strawberries	56·25	70·0	70·0	124
Raspberries	56·25	90·0	80·0	142
Blackcurrants	37·5	50·0	40·0	107
Mixed and other soft fruit	50·0	80·0	60·0	120
		smd per sq ft		
Tomatoes in glasshouses	0·04375	0·035	0·03	69
Flowers under glass	0·05	0·035	0·03	60
		smd per sq yd		
Flowers in the open	0·025	0·02	0·05	200
Hardy nursery stock	0·025	0·01	0·01	40
Bulbs	0·025	0·02	0·02	80
Mushrooms	0·05	0·035	0·02	40

TABLE 52

Standard Man-days in 1962, 1965 and 1968

Livestock	1962	1965	1968	1968 as a
		smd per head		percentage of 1962
Dairy cows	14·375	12·0	10·0	70
Dairy heifers in calf	3·75	2·5	3·5	93
Other female dairy cattle	3·125	2·5	2·5	80
Dairy bulls and bull calves	4·5	6·0	5·0	111
Beef cows	6·25	4·0	3·0	48
Beef heifers	3·75	2·5	2·5	67
Other beef cows	3·125	4·0	3·0	96
Beef bulls and bull calves	3·25	6·0	5·0	154
Other beef cattle	3·125	2·5	2·5	80
Sheep over 1 year	0·5	0·5	Hill 0·4	
			Other 0·7	
Breeding sows	4·375	4·0	4·0	91
Boars for service	2·5	4·0	4·0	160
Other pigs 2 months and over	2·5	1·0	1·0	40
Laying flock (pullets and hens)	0·3125	0·2	0·1	32
Pullets reared for laying flock	0·075	0·15	0·05	67
Other fowl	0·075	0·15	0·15	200
Broilers	0·0375	0·04	0·05	133
Turkeys	0·25	0·15	0·2	80
Ducks	0·25	0·15	0·1	40

Source: Department of Agriculture and Fisheries for Scotland

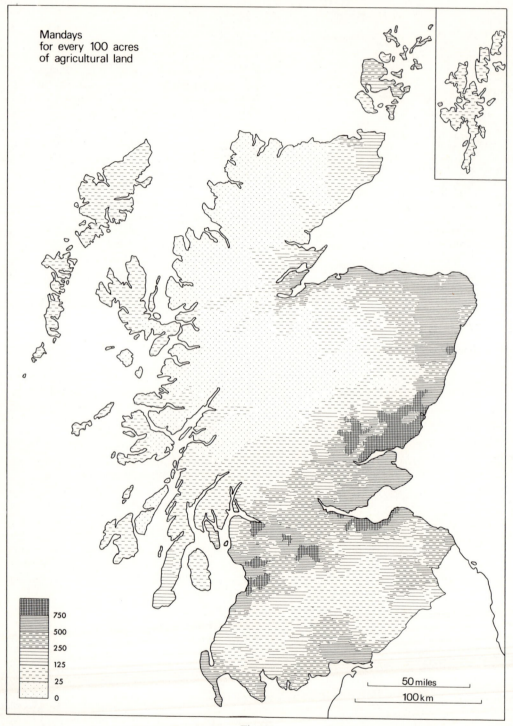

Mandays
for every 100 acres
of agricultural land

750
500
250
125
25
0

50 miles
100 km

Fig 191

MAN-DAYS

Number in 1965: 28,303,830. Number in 1970: 24,156,421

Smd per 100 acres of agricultural land: Mean 336; SD 229; Max 3,566; Min 7

While the levels at which the standard labour requirements have been fixed may be somewhat arbitrary, there is no doubt that the mapping of standard labour requirements per 100 acres (40ha) of agricultural land does give a valuable indication of the relative importance of agriculture throughout the country (Fig 191). In any case, for the present there is no satisfactory alternative. Of course, the figures, to which 15 per cent have been added to cover overheads, are dependent on the items included in the June census and no estimate has been made, for example, of the contribution of top fruit, which is enumerated in the December census; as on all the maps based on parish data, allowance must also be made for the large size and often linear shape of the upland parishes, as a result of which sharp contrasts on the ground, as along the Highland edge, are blurred on the map. Yet, whatever its limitations, Fig 191 is undoubtedly a more useful map than that of total labour (Fig 17), for the latter excludes the labour of the occupier and his wife and requires arbitrary judgments on the equivalence of part-time and casual labour.

The highest values recorded in 1965, exceeding 750 man-days (the equivalent of three full-time labour units per year) per 100 acres, were found in areas which depended heavily on horticultural crops, such as the Lothians to the east and west of Edinburgh, the Clyde valley south of Glasgow and the lowlands between Perth and Montrose; the last, which includes land on both sides of the Sidlaws, in Strathmore and along the coast from Dundee north-eastwards is by far the most important area. Such values were also found, though over a much smaller area, in the heart of the Ayrshire dairying district. Values of between 500 and 750smd per 100 acres, ie between 2 and 3 full-time labour equivalents per year, were found in the other principal cropping areas, the coastlands of the Moray Firth, lowland Kincardineshire, Fife and the remainder of the Lothians; in the Merse, by contrast, with its greater emphasis on cereals and sheep, only a few parishes were in this category. Values of between 500 and 750smd per 100 acres were also characteristic of the north-east (especially Buchan) where they were partly a reflection of the widespread occurrence of small, intensively worked family farms, and in the principal dairy areas in Ayrshire and the south-west. Over most of the remainder of the lowlands there were between 250 and 500smd per 100 acres, ie between 1 and 2 full-time labour equivalents throughout the year.

By contrast, values over most of the uplands were below 250smd, though there were differences between the Southern Uplands, with more than 25smd, and the Highlands, with fewer, the chief exception being the eastern and southern margins of the Highlands (though allowance must be made here for the inadequacies of the parish as a mapping unit). Values were also somewhat higher in the islands, especially in Orkney, which was unique among the crofting counties in having a number of parishes in which the number of smd per 100 acres exceeded 250. Of course, the downward revision of standard factors in 1968 will have had the effect of lowering these values by about 15–20 per cent. Subsequent maps, showing the relative importance of the major enterprises, are less likely to be affected by these changes.

ENTERPRISES

The term 'enterprise' requires some further qualification. All productive activities on a farm, such as the growing of a wheat crop, the breeding of pigs for sale or the keeping of a flock of laying fowl, can be regarded as an individual enterprise, and it may be desirable in some circumstances to treat each crop or class of livestock as a single enterprise. More commonly, however, it is convenient to group related activities, such as cereals or dairy cattle, which have common characteristics, either because they have similar requirements and to some extent are alternatives or because they are complementary. In any case, some items recorded in the census are clearly ancillary to others and it would generally be more useful to regard, say, bulls as part of cattle breeding rather than as a separate enterprise; similarly, fodder crops are not generally produced as ends in themselves. The main object of grouping is to facilitate comprehension and it is therefore desirable to reduce the number of enterprises recognised to manageable proportions. There are no definitive rules that can be prescribed a priori for doing so, but while a large number of enterprises will more closely approach reality, the larger the number the more difficult it will be to obtain any general picture and, given the nature of the census items on which grouping depends, the more frequently arbitrary decisions will have to be made. Thus, while having a single enterprise 'sheep' will avoid difficulties of allocating sheep between breeding and fattening, there is a loss of information, especially where, as with the various stages in the production of sheep or beef cattle, these related activities have distinctive distributions. The exact nature of the grouping of enterprises adopted will depend on the purpose for which it is required. That adopted here is intended for general use and seven major enterprises are recognised, viz dairying, beef cattle, sheep, pigs, poultry, cropping and horticulture. Grass is allocated to cattle and sheep according to the proportion of grazing units represented by each class of stock. Two maps showing further groupings of enterprises are also included, viz cattle and livestock, and one of a subdivision of an enterprise, cash cropping, which has not previously been analysed

in the preceding sections, of which the present section is to some extent a summary.

LIVESTOCK MAN-DAYS

Livestock smd per 100smd: Mean 67·2; SD 19·9; Max 99·9; Min 8·4

Livestock, including any grazing attributed to sheep and cattle, account for the majority of standard labour requirements on Scottish farms. In these and subsequent maps showing the relative importance of different enterprises, the distribution of man-days must be borne clearly in mind (Fig 191), for absolute values vary greatly and the actual numbers represented by a high percentage of all man-days may, in fact, be quite small. Thus, on Fig 192 the highest values occur in areas of extensive farming, those in the Highlands having less than 25smd per 100 acres. In many of these areas, where livestock accounted for more than 90 per cent of all labour requirements, no other enterprises were possible. Throughout western Scotland the proportion rarely fell below 80 per cent, the chief exceptions being the coasts of Ayrshire and Wigtown, with their early potato production, and the Clyde valley, where horticulture is locally important. Proportions were also high in the islands, although Orkney was exceptional in that livestock accounted for between 60 and 79 per cent of all smd in every parish but one. The lowest values, under 40 per cent, were found in the four principal areas for crop production, viz the lowlands between Perth and Stonehaven, the eastern half of Fife, the Lothians and the Merse, with somewhat higher values in the surrounding upland margins. In most of the north-east and on the coastal lowlands around the Moray Firth proportions were between 40 and 59 per cent.

CASH CROP MAN-DAYS

Cash crop smd per 100smd: Mean 13·9; SD 13·8; Max 98·0; Min 0·0

The definition of cash crops on the basis of census data is somewhat arbitrary; for present purposes, wheat, potatoes and sugar beet, together with half the barley acreage, were considered as cash crops, horticultural crops being

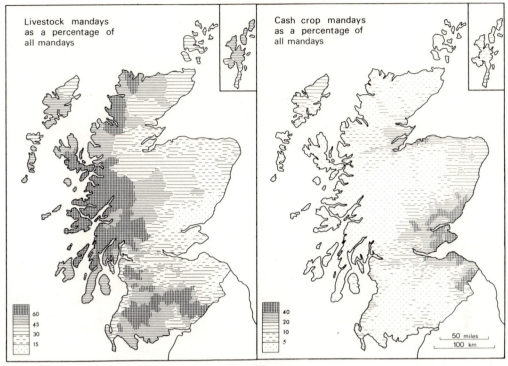

Livestock mandays
as a percentage of
all mandays

	60
	45
	30
	15

Cash crop mandays
as a percentage of
all mandays

	40
	20
	10
	5

50 miles
100 km

Figs 192–193

treated separately (Fig 201). Fig 193 is, to a large extent, the mirror image of Fig 192, for proportions of smd attributable to cash crops were highest in Strathmore, the lowlands between Dundee and Montrose, east Fife, the Lothians and the Merse; apart from narrow strips of land around the Moray Firth and along the coasts of Ayr, most other parishes with 10 per cent or more of smd from cash crops were in eastern Scotland, with values between 10 and 20 per cent over most of the north-east. In the Highlands and Southern Uplands, by contrast, the proportions were all below 5 per cent and no cash crops at all were grown in a large number of parishes.

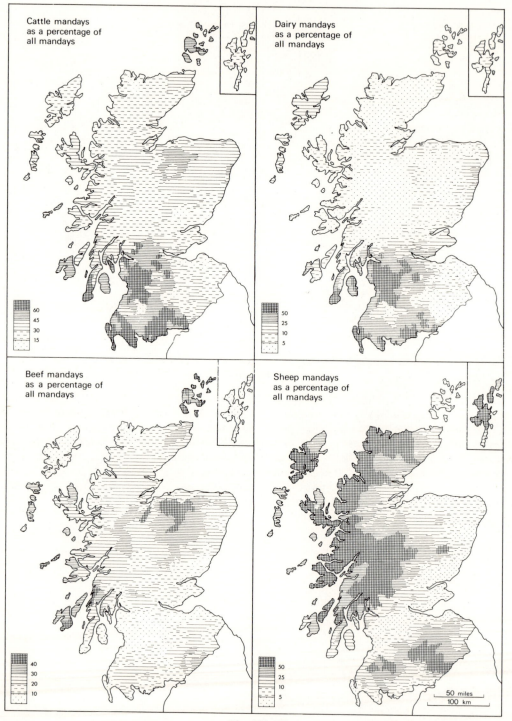

Figs 194–197

CATTLE MAN-DAYS

Cattle smd per 100smd: Mean 38·6; SD 16·5; Max 80·7; Min 4·8

Cattle were relatively most important in relation to other enterprises in the lowlands of the south and west, chiefly in Ayrshire, Galloway and Dumfriesshire (Fig 194). Throughout the uplands cattle accounted for less than 30 per cent of often small totals, and in north-east Scotland for between 30 and 60 per cent, with a small outlier in Orkney where values exceeded 45 per cent. In eastern Scotland, too, proportions were generally under 30 per cent and fell below 15 per cent in a few parishes where horticulture was important.

DAIRY MAN-DAYS

Dairy smd per 100smd: Mean 18·2; SD 19·0; Max 73·8; Min 0·0

Although dairy cattle are much more localised, the general pattern of relative importance, as revealed by Fig 195, was very similar to that for all cattle, with values exceeding 50 per cent in much the same areas. However, with the exception of south Fife, and the environs of Aberdeen, there were few other areas where dairying was relatively important as an enterprise, and in both the uplands and the main arable areas dairying accounted for less than 5 per cent.

BEEF MAN-DAYS

Beef smd per 100smd: Mean 20·5; SD 10·6; Max 56·6; Min 0·7

The main areas where beef cattle represented a relatively important enterprise were in north-east Scotland (Fig 196), although the proportions were lower than might have been expected from a consideration of the beef herd's contribution to gross output. Proportions were much more uniform throughout the country than were those of dairy cattle, percentages below 10 per cent being mainly confined to the western half of the central lowlands.

SHEEP MAN-DAYS

Sheep smd per 100smd: Mean 22·4; SD 21·0; Max 93·9; Min 0·0

Sheep, by contrast, tended to show a much greater range of values. Sheep accounted for half or more of all smd in parts of the Southern Uplands and over most of the western Highlands, with percentages of 25 or more in most of the remaining uplands except the north-east (Fig 197). Values were below 5 per cent in most of the central lowlands and in the principal areas for crop production, except the Merse, where proportions of 10 per cent and over were recorded. Orkney was again anomalous, having relatively few sheep in relation to other enterprises.

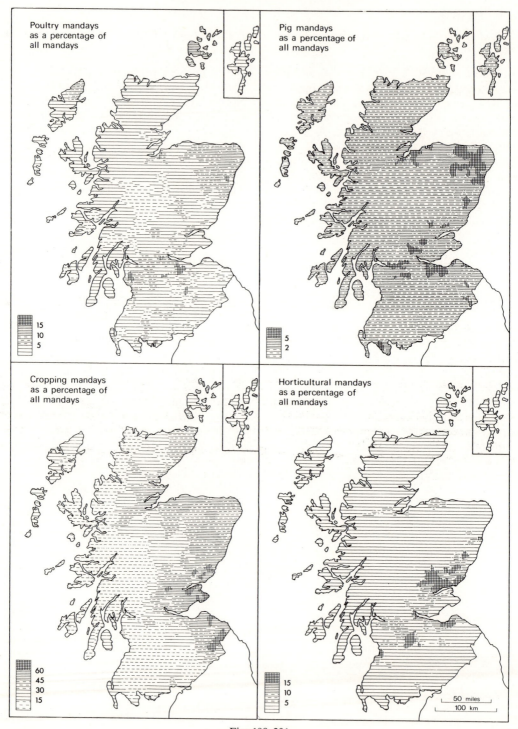

Figs 198–201

POULTRY MAN-DAYS

Poultry smd per 100smd: Mean 4·3; SD 4·5; Max 38·6; Min 0·3

Poultry represent a relatively minor enterprise, accounting for 15 per cent or more of all smd in only a few parishes, mainly in the central lowlands (Fig 198). Poultry were also relatively important in two other areas, north-east Scotland and Orkney, though their contribution even there was small.

PIG MAN-DAYS

Pig smd per 100smd: Mean 1·9; SD 2·5; Max 24·6; Min 0·0

The distribution of parishes in which pigs were a relatively important enterprise is very sporadic and their contribution was even smaller than that of poultry, with only a few parishes in which pigs accounted for 5 per cent or more of all smd (Fig 199). Pigs were relatively more important in the central lowlands and in north-east Scotland.

CROPPING MAN-DAYS

Cropping smd per 100smd: Mean 29·0; SD 16·8; Max 69·3; Min 0·2

Cropping differs from the cash crop enterprise shown in Fig 193 in that all crops other than horticultural crops are included (Fig 200). The pattern of the two maps is, however, fairly similar, with values of 60 per cent and over in the lowlands north-east of Dundee, in eastern Fife and the Merse, and 45 per cent and over in the remaining areas which are important for crop production. Even outside these cropping areas in other parts of eastern Scotland 30 per cent or more of all smd were devoted to cropping (mainly of fodder crops), whereas less than 15 per cent were devoted to cropping in most western parishes.

HORTICULTURAL MAN-DAYS

Horticultural smd per 100smd: Mean 3·8; SD 91·4; Max 78·9; Min 0·0

Horticulture is the most highly localised of all enterprises mapped in this section and the decision on what constitutes a horticultural crop is somewhat arbitrary, for some crops, such as peas for canning, drying or freezing, are essentially farm crops grown on a field scale, others, such as raspberries, are grown by both farmers and specialist market gardeners, while others, such as nursery stock, are largely the province of specialists. For present purposes, horticulture embraces all those items included as horticultural crops in the agricultural census and shown as such in Table 52.

Parishes in which horticultural crops were relatively important were generally also those in which large acreages were grown, notably in Strathmore and the lowlands from Dundee to Montrose, East Lothian and the Clyde valley (Fig 201). Outside the central lowlands and their eastern extension, horticultural crops were of very minor importance.

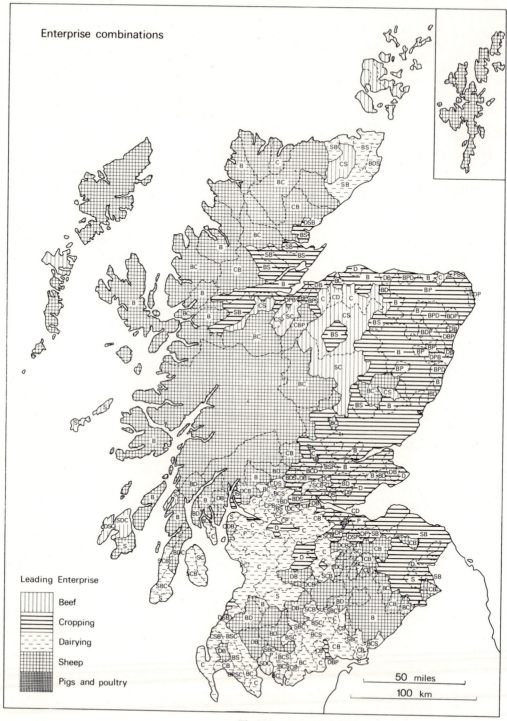

Fig 202

Enterprise combinations

Number of enterprises
5
4
3
2
1

50 miles
100 km

Fig 203

ENTERPRISE COMBINATIONS

Like the preceding maps of crop and livestock combinations (Figs 106 and 189), Fig 202 is an attempt to synthesise the information shown in Figs 192 to 201 on a single map; the various shadings indicate the first-ranking (leading) enterprise in each parish and the letters identify in rank order the other components in each enterprise combination; other enterprises may, of course, be present, but not in sufficient strength to be recognised by the method of least squares used to identify enterprise combinations. The letters represent the following enterprise: B, beef cattle; C, cropping; D, dairying; P, pigs and poultry; and S, sheep. Horticultural crops are included with cropping; pigs and poultry are treated as a single enterprise; grass smd are allocated to cattle and sheep according to their share of grazing units; and all proportions are calculated in relation to total man-days before the addition of 15 per cent to cover overheads.

The pattern which emerges from this map repeats in broad outline what has already been shown in the preceding maps in this section; for,

in general, enterprises achieved leading rank in those parishes in which they were most important. There are, however, two anomalies. The area in which cropping was dominant is much larger than might have been expected on other grounds. This is partly because horticultural crops have been included in cropping and also because fodder crops are allocated to cropping rather than to the live-stock which they support (as has been done with grass). This latter procedure particularly affects the extent of the area dominated by beef cattle, for dairying is concentrated in south-east Scotland, where relatively few crops are grown and dairy cattle depend heavily on purchased feed, whereas beef cattle are most prominent in the north-east, where large acreages of fodder crops are grown. Sheep, like dairy cattle, are relatively little affected, since large numbers are kept on rough grazings. The importance of beef cattle is thus considerably understated, though it was the second ranking enterprise after crops in most parishes in north-east Scotland. This difficulty is largely avoided in the succeeding section, where the farm is treated as the unit and empirical rules have been devised to allocate holdings to particular types of farm. Beef was also the second-ranking enterprise in most of the upland parishes in which sheep were the leading enterprise. In the main areas of crop production, crops were often sufficiently important to be the sole enterprise identified; but in the Merse, sheep frequently appeared as a second-ranking enterprise. In the parishes in which dairying was dominant, the second-ranking enterprise was often cropping, though the acreages under crops were generally small.

This map is complex and is intended for detailed examination. The broad pattern is revealed by the shadings attributed to the leading enterprises and by Fig 203, which shows the number of enterprises identified in the enterprise combination in each parish. Rather unexpectedly the only large areas in which a single enterprise is identified lie between Perth and Montrose, in the western half of the central lowlands and in the western Highlands, though a single enterprise would probably have been recorded over a much larger area of upland if parishes had been smaller and included less low ground. These areas apart, there is some tendency for numbers of enterprises to be highest in the central lowlands, in north-east Scotland and in the south-west.

6

Types of Farm

The enterprises discussed in the preceding section are rarely the sole enterprise on farms (though there is a long-term trend for each enterprise to be undertaken on fewer holdings and for each holding to have fewer enterprises). Rather, they are components of farming systems which are the result of each farmer's assessment of the most appropriate enterprises for his farm in the light of its size and resources, both natural and man-made, his judgments on the relative profitability of various enterprises and his preferences.

These farming systems are not self-evident. Although such terms as 'dairy farm' and 'market garden' are commonly used in every-day speech, they are not very precise, for the mix of enterprises shows great variation; for most enterprises, there is a continuum from those holdings where there is only one enterprise to those where it makes only a token appearance. To assist the making of policy and the provision of advice to farmers, the Department of Agriculture and Fisheries has attempted from time to time to classify farms objectively on the basis of census data. This section is based on the Department's work.

This dependence on the census imposes certain limitations on classification in that only information on the areas under different land uses and on numbers of livestock is available and then only in respect of one particular day in the year. The approach which is now adopted is to convert all relevant census data into labour requirements on the basis of standard man-days (smd) and to devise objective rules by which these data can be used to identify farming types. The procedure described in this section is that used with data from the 1962 census and with slight modification since, and applies only to those holdings which, on the basis of imputed man-days, are classified as full-time, ie those with 250 or more smd. It will, however, be recalled (Section 2) that there is in fact a 'grey area' around this boundary, for some holdings which are nominally full-time are in fact run on a part-time basis while other holdings with fewer than 250smd are in fact full-time. There have also been several changes since 1962, particularly in respect of the standard labour requirements used, which have been revised downwards so that the total number of man-days in 1970 was only 70 per cent of that in 1962. There was also a major revision of census procedure in 1970 when 16,034 holdings with fewer than 26smd were eliminated from the census as statistically insignificant; these holdings occupied 955,005 acres (386,485ha) of agricultural land, though only 5,456 acres (2,208ha) were under crops and by far the greater part was rough grazing.

The total number of smd is first computed for each holding and 15 per cent added to cover overheads, including such tasks as maintenance which cannot be attributed to any single enterprise, though these overheads are not considered in computing the ratios and percentages under different enterprises. Those holdings with 250smd or more are then subdivided according to prescribed rules and in a defined order, since a holding may fall into more than one category; there is also a residual category for those holdings which are not otherwise classified. To simplify description the following terms are used:

total man-days: all smd except those attributed to overheads

net man-days: total smd less those devoted to pigs and poultry

grazing ratio: percentage of total area in rough grazing and unmown grass 7-years old and over

intensive ratio: percentage of total smd devoted to horticulture, pigs and poultry

crop ratio: percentage of net smd devoted to crops (including horticulture)

sale crop ratio: percentage of net smd in wheat, barley, potatoes, sugar beet and horticulture

Eight major types of farm are recognised, some of which are further subdivided, though these subdivisions are not considered in this atlas. These types were:

INTENSIVE

The first types to be recognised were horticultural, pig and poultry farms, all sub-types of intensive farms; allocation to the fourth sub-type, mixed intensive farms, was made after the identification of dairy farms. The first three sub-types had 60 per cent or more of all smd devoted to the appropriate enterprise; farms in the last sub-type had 70 per cent or more in dairy cows, horticulture, pigs and poultry, but those where the percentage in pigs, poultry and horticulture was less than 70 per cent and the holding was not registered as a milk-selling farm were excluded. This definition has since been modified to include those holdings with an intensive ratio of 60 per cent or more.

DAIRY

These were farms registered as milk-selling farms, provided that they had at least four dairy cows in milk and that at least 25 per cent of net smd were devoted to dairy cows; farms with 20 per cent or more could be considered, if 50 per cent of net smd were in dairy cows and sheep. Farms to be classified as mixed intensive were excluded even if they satisfied these conditions.

HILL SHEEP

Hill sheep farms were those where the grazing ratio was at least 90 per cent, the crop ratio less than 15 per cent and at least 35 per cent of total smd were devoted to sheep. Farms where cattle accounted for a higher proportion of smd than did sheep or where the intensive ratio was 25 per cent or more were excluded.

CROPPING

Cropping farms were those where the crop ratio was 75 per cent or more, or where the crop ratio was at least 50 per cent and the sale crop ratio at least 25 per cent, except holdings on which pigs and poultry accounted for 50 per cent or more of total smd or where the grazing ratio was 60 per cent or more or the occupier had a share of common grazings.

UPLAND

Upland farms were those where the crop ratio was less than 30 per cent, but higher ratios could be accepted if they were accompanied by higher grazing ratios, eg those with a crop ratio of 55 per cent or over and a grazing ratio of 75 per cent or more. Farms with a sale ratio of 15 per cent or more were excluded, as were those where the ratio of cattle smd to sheep smd was under 30 per cent or the intensive ratio was 25 per cent or more.

REARING WITH INTENSIVE LIVESTOCK

Rearing with intensive livestock farms were those of the remainder which had an intensive ratio of 25 per cent or more.

ARABLE REARING AND FEEDING

Arable rearing and feeding farms were those among the remainder where the crop ratio was 55 per cent or over, though farms with a crop ratio of at least 45 per cent and a sale ratio of at least 25 per cent were also included. Farms with a grazing ratio of 60 per cent or more or where the occupier had a share of common grazings were excluded.

REARING WITH ARABLE

These were the remaining farms, ie those with a crop ratio of less than 55 per cent, a lower grazing ratio than that required for any given crop ratio on upland farms, an intensive ratio of less than 25 per cent and a sale crop ratio of less than 25 per cent.

PART-TIME AND SPARE-TIME

Holdings with fewer than 250smd were not classified according to type, but were divided into part-time holdings, ie those with between 100 and 250smd, and spare-time, ie those with fewer than 100smd.

REGIONAL DISTRIBUTION OF FARM TYPES

Once farms have been classified according to type, their distribution can be plotted and their relative importance assessed. Somewhat different pictures will, however, emerge depending on the basis of assessment, as has already been shown in the discussion on size of holding (Figs 39–46). The use of numbers as the basis of evaluation emphasises the number of decision makers, irrespective of the size of the units which they control; the employment of area as the criterion draws attention to the contribution of the different groups to the landscape and ignores the fact that a large but extensively run holding may produce less than an intensively worked small holding; and reference to standard labour requirements provides one measure of the size of business. As Table 53 shows, the main difference is between the percentages of holdings according to acreage and those according to numbers, the employment of labour require-

ments producing an intermediate result since the proportions of part- and spare-time holdings are similar to those calculated on an acreage basis and the proportions of hill sheep and upland farms resemble those based upon numbers. As has been noted, results from the 1962 census and those from the 1970 census are not strictly comparable since large numbers of small and part-time holdings were removed from the census in 1970, but the distribution of percentages between the different types of farm is broadly similar in both years. The major exception is that upland farms were relatively much more important in 1970 than in 1962, having increased by 34 per cent, and rearing with arable farms, the number of which was only 59 per cent of the 1962 total, correspondingly few; it is clear that some rearing with arable farms had been reclassified as upland, and, if the two classes are treated together, differences between the two years are quite small. Numbers of both part- and spare-time holdings have fallen more markedly than those of full-time (viz to 50 and 49 per cent of 1962 numbers respectively, compared with 78 per cent), but this decline is largely due to the exclusion of agriculturally insignificant holdings in 1970. The other two categories to lose ground relatively are both classes of mixed farms, viz rearing with intensive livestock and arable rearing and feeding (which is the residual category).

The contribution of the different types of holdings also varies regionally, for one of the

TABLE 53

Types of Farms in Scotland in 1962 and 1970

	Hill sheep	Upland	Rearing with arable	Rearing with intensive livestock	Arable rearing and feeding	Cropping	Dairy	Intensive	Part-time	Spare-time
				Percentage of farms by number						
1962	2	6	9	3	4	7	12	3	19	34
1970	3	13	9	2	3	10	14	4	15	27
				Percentage of farms by area						
1962	37	19	10	1	3	6	11	—	6	8
1970	37	26	6	1	2	7	9	1	7	5
				Percentage of farms by smd						
1962	4	10	15	3	6	18	32	7	4	2
1970	4	16	10	2	4	20	27	11	4	2

Source: *Scottish Agricultural Economics*, Vol 15, and *Agricultural Statistics 1970*

most striking features of the agricultural geography of Scotland is the extent to which particular types of farms are localised in different parts of the country. This is even revealed by Table 54, which shows proportions by type of farm according to acreage and mandays. These figures are calculated as a proportion of full-time farms only and, with the exception of the changed importance of upland and rearing with arable farms, show an even closer similarity between the results for the two years. Both by area and by standard labour requirements the Highlands were dominated by hill sheep and upland farms, but in the remaining regions there was a contrast between the results by area and those by smd. By area, hill sheep and upland farms were always important, accounting for half or more of total acreage except in the North East, where there were very few hill sheep farms; by standard labour requirements, hill sheep farms were everywhere of minor importance, and the more intensive farming types, notably dairy farms in the South West and cropping farms

in the East Central Region, accounted for the majority of smd in each region.

These types do not represent well-defined, discrete groups which can be clearly separated, and there is a great range of conditions within any one type, both locally and regionally. The boundaries separating groups are better regarded as points on a continuum, and the different types of farm can be ranked in the order of importance of their major components. In general, in the progression from upland to lowland, systems become more diverse, the proportion of rough grazing decreases and the ratio of cattle to sheep increases; but there are obvious regional differences in respect of both the relative importance of enterprises within any type of farm (as in the contrast between the role of sheep in the South East and in other Regions) and the size of enterprise. Tables 55–61 give some indication of the range of variation.

As has already been noted in Section 3, there are considerable variations in average farm size and these are manifest both within any type of

TABLE 54

Types of Full-time Farms by Regions in 1962 and 1970

	Hill sheep	Upland	Rearing with arable	Rearing with intensive livestock	Arable rearing and feeding	Cropping	Dairy	Intensive
			Percentage of full-time farms by area					
Highland 62	67	24	4	—	1	1	3	—
Highland 70	66	26	2	1	—	1	3	—
North East 62	8	17	39	6	12	9	8	1
North East 70	6	45	19	3	7	11	7	1
East Central 62	37	20	11	—	3	21	7	1
East Central 70	38	24	6	1	2	23	5	1
South East 62	29	21	20	1	4	16	8	1
South East 70	31	28	12	1	4	18	5	1
South West 62	33	22	3	1	1	1	39	9
South West 70	34	30	3	1	1	1	30	1
			Percentage of full-time farms by smd					
Highland 62	22	24	16	1	8	7	20	2
Highland 70	24	27	10	1	5	12	18	3
North East 62	—	8	34	10	15	13	16	3
North East 70	—	19	25	6	9	17	16	8
East Central 62	3	6	9	1	5	47	18	11
East Central 70	3	7	5	1	3	48	14	20
South East 62	5	11	20	1	6	30	17	9
South East 70	5	18	15	2	5	31	12	12
South West 62	4	11	3	1	1	2	69	9
South West 70	4	20	3	1	1	3	61	8

Source: *Scottish Agricultural Economics*, Vol 15, and *Agricultural Statistics 1970*

farm and between regions. As Table 55 shows, hill sheep farms were approximately ten times as large (by area) as the average for all full-time farms and had the largest average area in all regions; but hill sheep farms in the Highlands were almost three times as large as those in southern Scotland. The large change in the average for such farms in the North East between 1962 and 1970 is probably a reflection of the very small number of farms in this class in that region. Upland farms were also above average in area and were again largest in the Highlands, though the range of differences was small. Mainly because of the high proportion of hill sheep and upland farms in the Highlands and their large size, the average area of all full-time farms was also highest in this region, being more than three times as large as the Scottish average. Among the lowland and semi-upland types, farms tended to be largest in the South East, which had the second largest average area of farm, and smallest in the North East. The figures for the two years are broadly similar, though there is widespread evidence of increases in average farm size, especially among the more intensive types.

The average size as measured by smd shows a much smaller range of values, both as between regions and within types (Table 56). On this basis, farms were smallest in the North East and in the Highlands, and largest in the East Central and South East Regions, and among cropping, dairy and intensive farms. The figures for 1962 and 1970 are not strictly comparable because of the downward revision of the factors used in calculating smd; but the general pattern of values is similar in both years.

There are similar differences in the major components. Table 57 shows regional differences in the average area of land under tillage on the different types of farm, though regional differences in farm size must be borne in mind in interpreting these and subsequent tables. The declining size of enterprise with type may also be somewhat misleading in that not all farms have the enterprise in question and the proportion having it is likely to decline with the importance of that enterprise. If rearing with intensive livestock farms, which are generally small, are disregarded, there is a progressive increase in average area from hill sheep to cropping farms; averages were generally smaller

TABLE 55

Average Size of Farm (by area) in 1962 and 1970

	Highland	North East	East Central	South East	South West	Scotland
			Acres in 1962			
Hill sheep	6,086	6,201	4,318	2,157	2,150	4,093
Upland	1,520	359	1,030	825	489	774
Rearing with arable	429	195	414	531	223	270
Rearing with intensive livestock	187	74	122	234	100	86
Arable rearing and feeding	195	135	232	335	194	168
Cropping	245	173	224	339	170	229
Dairy	357	189	226	223	224	228
Intensive	54	52	42	45	28	36
Part-time	101	48	51	32	43	73
Spare-time	63	91	70	19	30	62
Full-time	1,895	191	457	501	347	478
			Acres in 1970			
Hill sheep	6,548	4,666	4,268	2,278	2,238	4,182
Upland	1,776	547	1,005	755	456	747
Rearing with arable	384	170	424	498	270	248
Rearing with intensive livestock	1,352	133	195	349	183	207
Arable rearing and feeding	233	171	291	347	239	216
Cropping	316	221	271	373	189	273
Dairy	506	245	266	225	224	247
Intensive	53	75	44	60	22	45
Part-time	350	70	106	99	100	162
Spare-time	84	40	30	43	43	66
Full-time	2,301	282	540	569	385	576

Source: *Scottish Agricultural Economics*, Vol 15, and *Agricultural Statistics 1970*

TABLE 56

Average Size of Farm (by smd) in 1962 and 1970

	Highland	North East	East Central	South East	South West	Scotland
		Smd in 1962				
Hill sheep	936	597	1,069	1,036	912	953
Upland	705	778	956	1,367	909	882
Rearing with arable	746	751	1,145	1,643	863	879
Rearing with intensive livestock	878	621	990	1,050	554	651
Arable rearing and feeding	948	780	1,367	1,446	1,016	916
Cropping	1,428	1,128	1,648	1,995	1,094	1,541
Dairy	1,081	1,635	1,986	1,541	1,445	1,498
Intensive	899	1,219	1,400	1,137	1,222	1,249
Part-time	115	140	143	144	138	128
Spare-time	25	29	34	29	33	27
Full-time	881	861	1,494	1,544	12,57	1,152
		Smd in 1970				
Hill sheep	819	614	870	828	742	799
Upland	646	646	770	1,111	744	738
Rearing with arable	677	611	884	1,325	806	727
Rearing with intensive livestock	924	783	1.127	1,615	800	878
Arable rearing and feeding	857	637	1,135	1,298	838	789
Cropping	1,194	973	1,475	1,448	915	1,300
Dairy	935	1,495	1,762	1,239	1,183	1,251
Intensive	1,189	1,594	2,354	1,400	1,061	1,591
Part-time	155	173	172	170	169	167
Spare-time	51	56	56	55	53	53
Full-time	8,025	796	1,433	1,276	1,005	1,025

Source: DAFS and *Scottish Agricultural Economics*, Vol 15, and *Agricultural Statistics 1970*

TABLE 57

Average Area under Tillage in 1962 and 1970

	Highland	North East	East Central	South East	South West	Scotland
		Acres in 1962				
Hill sheep	7	2	8	8	6	7
Upland	18	31	28	49	21	26
Rearing with arable	44	45	62	101	44	52
Rearing with intensive livestock	28	28	40	55	13	28
Arable rearing and feeding	71	61	83	118	61	69
Cropping	107	89	120	189	77	121
Dairy	33	70	82	65	33	43
		Acres in 1970				
Hill sheep	5	6	7	7	4	5
Upland	19	30	26	50	17	25
Rearing with arable	59	55	72	132	62	65
Rearing with intensive livestock	26	54	74	130	23	58
Arable rearing and feeding	99	80	112	193	92	97
Cropping	177	126	168	248	102	168
Dairy	29	84	105	63	29	43

Source: *Scottish Agricultural Economics*, Vol 15, and *Agricultural Statistics 1970*

in the North East, where farms are small, and in the South West, where conditions are generally unfavourable to tillage; they thus tended to be low on the dairy farms which predominate in the latter Region.

Beef cattle are the most important enterprise on Scottish farms and the average size of herd of breeding cows was highest on the hill sheep and upland and rearing with arable farms, and smallest on the lowland types, especially dairy farms (Table 58). Apart from a tendency for herds in the North East and, to a lesser extent, in the South West Regions to be somewhat smaller, there was little regional variation in herd size. The large increases in average herd size between 1962 and 1970 reflect both farm enlargement and the increasing importance of beef breeding.

Numbers of other beef cattle were more uniformly distributed and tended to be higher both on upland and rearing farms and on cropping farms, and to be smallest on such different types as hill sheep and dairy farms (Table 59). Regionally, such cattle were most important in eastern Scotland. However, as has been shown in Section 5, the category 'other beef cattle' includes both animals being reared and those being fattened, and there were quite notable differences in their distribution (Table 60). The largest herds of younger cattle were

found on upland farms and the smallest on dairy farms and cropping farms, and herds were generally largest in the East Central and South East Regions. The largest herds of older beef cattle were on arable rearing and feeding farms and the smallest on hill sheep farms.

Table 61 shows the average size of ewe flock in 1962 and 1970, though it must be interpreted with care in view of the absence of breeding flocks from many intensively-run lowland farms. As is to be expected, flocks were largest on hill sheep farms, but there were considerable regional differences, with the smallest flocks in the North East Region, a characteristic feature of many other types, too. The largest flocks tended to occur in the South East Region for nearly all types.

FARM TYPE AND GROSS OUTPUT

There is one other way in which the character of the different types of farms can be considered, the composition of their gross output, although this can be examined only on the basis of sample figures for the whole of Scotland. Table 62 shows the main groups of products as a proportion of gross output in 1962/3 and 1970/1; the class Livestock with Arable in 1962 is a combination of the Arable Rearing and Feeding and Rearing with Arable farms described elsewhere in this Section. Figures for cattle include

TABLE 58

Average Number of Beef Cows in 1962 and 1970

	Highland	North East	East Central	South East	South West	Scotland
		Number in 1962				
Hill sheep	15	7	18	16	12	15
Upland	25	27	31	35	23	26
Rearing with arable	19	13	24	28	7	16
Rearing with intensive livestock	15	5	9	9	4	5
Arable rearing and feeding	14	5	21	10	6	8
Cropping	15	4	9	7	3	7
Dairy	3	1	1	1	1	1
		Number in 1970				
Hill sheep	22	20	26	24	22	23
Upland	43	41	48	50	41	43
Rearing with arable	26	16	34	37	17	20
Rearing with intensive livestock	53	8	16	14	13	12
Arable rearing and feeding	22	6	27	14	7	11
Cropping	10	4	9	7	3	7
Dairy	5	2	1	2	4	3

Source: *Scottish Agricultural Economics*, Vol 15, and *Agricultural Statistics 1970*

TABLE 59
Average Number of Other Beef Cattle in 1962 and 1970

	Highland	North East	East Central	South East	South West	Scotland
			Number in 1962			
Hill sheep	19	10	23	22	20	20
Upland	35	50	51	57	56	49
Rearing with arable	31	45	49	57	47	45
Rearing with intensive livestock	21	31	26	25	19	29
Arable rearing and feeding	35	44	57	47	47	45
Cropping	37	42	44	43	32	43
Dairy	12	13	12	7	11	11
			Number in 1970			
Hill sheep	25	18	28	28	27	26
Upland	52	66	65	70	60	67
Rearing with arable	42	80	55	66	69	65
Rearing with intensive livestock	60	58	42	45	50	55
Arable rearing and feeding	50	56	72	83	68	61
Cropping	36	49	47	47	34	47
Dairy	17	24	14	10	21	20

Source: *Scottish Agricultural Economics*, Vol 15, and *Agricultural Statistics 1970*

TABLE 60
Average Number of Other Beef Cattle Under One-year Old and One-year Old and Over in 1970

	Highland	North East	East Central	South East	South West	Scotland
		Number of other beef cattle under one-year old				
Hill sheep	19	18	24	23	21	21
Upland	39	40	48	51	32	44
Rearing with arable	25	25	40	43	30	29
Rearing with intensive livestock	50	19	29	23	23	22
Arable rearing and feeding	22	13	38	40	23	20
Cropping	13	10	17	19	10	15
Dairy	12	15	9	10	13	12
		Number of other beef cattle one-year old and over				
Hill sheep	6	2	3	6	6	5
Upland	13	26	16	19	28	23
Rearing with arable	18	54	15	22	39	36
Rearing with intensive livestock	10	38	12	22	28	32
Arable rearing and feeding	28	43	34	43	45	41
Cropping	23	39	30	28	24	31
Dairy	6	10	5	5	8	8

Source: *Agricultural Statistics 1970*

TABLE 61
Average Number of Breeding Ewes in 1962 and 1970

	Highland	North East	East Central	South East	South West	Scotland
			Number in 1962			
Hill sheep	831	520	941	902	797	839
Upland	227	88	272	432	207	211
Rearing with arable	91	50	140	324	82	90
Rearing with intensive livestock	71	15	33	95	29	21
Arable rearing and feeding	75	29	78	189	61	52
Cropping	120	33	45	117	33	55
Dairy	95	23	31	59	50	49
			Number in 1970			
Hill sheep	931	557	988	930	827	898
Upland	256	90	262	416	189	196
Rearing with arable	94	38	125	274	84	77
Rearing with intensive livestock	95	17	31	140	45	33
Arable rearing and feeding	61	26	79	159	61	50
Cropping	60	27	30	63	22	36
Dairy	58	17	15	40	37	36

Source: *Agricultural Statistics 1970*

TABLE 62

Contributions of Livestock to Gross Output in 1962/3 and 1970/1

	Hill sheep	Upland	Livestock with arable	Rearing with intensive livestock	Cropping	Dairy
			Percentage of gross output 1962/3			
Cattle	14	39	38	28	20	72
Sheep	81	14	17	7	8	5
Crops	1	7	27	15	56	13
Pigs and poultry	—	3	10	44	12	6

	Hill sheep	Upland	Livestock with arable	Rearing with intensive livestock	Arable and feeding	Cropping	Dairy
			Percentage of gross output 1970/1				
Cattle	25	56	55	33	45	26	79
Sheep	73	25	12	1	4	5	3
Crops	1	5	25	8	33	54	7
Pigs and poultry	—	11	6	54	14	14	9

Source: *Scottish Agricultural Economics*, Vols 15 and 22

TABLE 63

Sales of Store and Fat Lambs in 1962/3

	Hill sheep	Upland	Livestock with arable	Rearing with intensive livestock	Cropping	Dairy
			Percentage of sales of all sheep			
Store lambs	62	46	33	37	11	16
Fat lambs	9	27	31	45	33	48
			Ratio of sales of store lambs to those of fat lambs			
	7·0	1·7	1·1	0·8	0·3	0·3

Source: *Scottish Agricultural Economics*, Vol 15

products from cattle (chiefly milk) as well as sales of cattle, although only on dairy farms, where the percentage was 59 per cent in 1962/3 and 62 per cent in 1970/1, did the contribution of milk exceed 1 per cent of total gross output; figures for sheep include wool, those for crops include horticulture and those for pigs and poultry include eggs. The figures confirm the picture provided by the preceding tables, particularly the increasing importance of crops in the progression from hill sheep to cropping farms, though the grouping of all sales of cattle

and those of sheep under single heads hides the parallel change from breeding and rearing to fattening. Some indication of this is given by Table 63, which records for 1962/3 the proportion of sales of sheep that were due to store and fat lambs respectively and the ratio between them; the contribution of cast ewes (17·3 per cent on hill sheep and 9·4 per cent on upland farms) and of fat sheep (34·6 per cent on cropping and 10·8 per cent on dairy farms) should be borne in mind in interpreting these figures. Similar relationships exist with store and fat cattle; thus, whereas on upland farms sales of cows and heifers accounted for 18·8 per cent, those of calves for 27·2 per cent and those of store cattle for 46·7 per cent, on cropping farms 12·8 per cent of sales came from store cattle and 83·3 per cent from fat cattle.

The distribution of the various types and their relative importance can now be considered. For each of the eight main types three maps are presented, a dot map showing the distribution of such farms in 1962 and two choropleth maps, the relative importance of the type in question being measured by its contribution to the acreages and standard man-days under all types. These maps and the comments on the composition of the different types of farm must be read in conjunction, for the text adjoining the maps has deliberately been kept short.

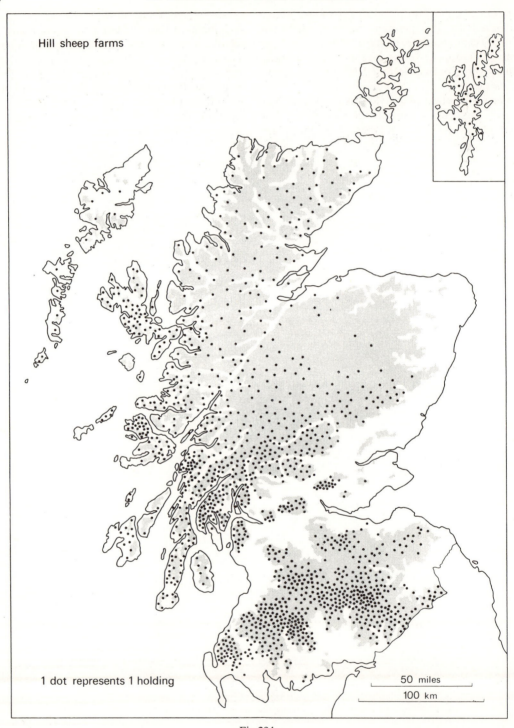

Hill sheep farms

1 dot represents 1 holding

50 miles

100 km

Fig 204

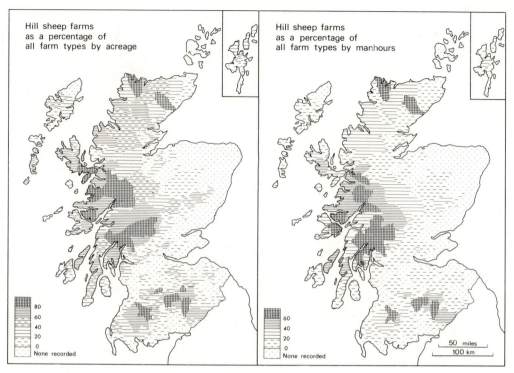

Figs 205–206

Hill Sheep Farms

Number: 1,424 in 1962; 1,275 in 1970

Percentage of full-time farms: 5·0 in 1962; 5·8 in 1970

Percentage of total area: 37·2 in 1962; 37·2 in 1970

Percentage of total smd: 3·9 in 1962; 4·2 in 1970

Hill sheep farms are, by definition, largely confined to land under rough grazing, but they are not co-extensive with it; for rather less than two-thirds of the rough grazing is on such farms. As Fig 204 shows, such farms were widespread in 1962, but the differences in farm size recorded in Table 55 must be borne in mind in interpreting this map; all dots are not of equal significance. Such farms were most numerous in the southern Highlands, in Argyll and Perthshire and in the Southern Uplands. Table 64 records the relative contribution of hill farms in the five regions; the large proportion of the agricultural land in such farms in the Highlands confirms both the larger size and the lower intensity of farms in this Region. Figs 205 and 206 provide a more precise indication of regional differences. By area, hill sheep farms were relatively most important in

TABLE 64

Regional Distribution of Hill Sheep Farms in 1962 and 1970

	High-land	North East	East Central	South East	South West	Scot-land
	Percentage by numbers					
1962	41	2	13	13	32	100
1970	37	2	14	14	32	100
	Percentage by total area					
1962	60	3	14	7	17	100
1970	58	2	15	8	17	100
	Percentage by total smd					
1962	40	1	14	15	30	100
1970	38	2	16	15	30	100

Source: *Scottish Agricultural Economics*, Vol 15, and *Agricultural Statistics 1970*

the western Highlands and in the western and central parts of the Southern Uplands, contributing 80 per cent or more in a number of parishes (Fig 205). Measured by smd, their contribution was less marked, though the same areas were still relatively most important (Fig 206). In interpreting these maps allowance must be made for the large size of most upland parishes and the fact that many will include some low ground.

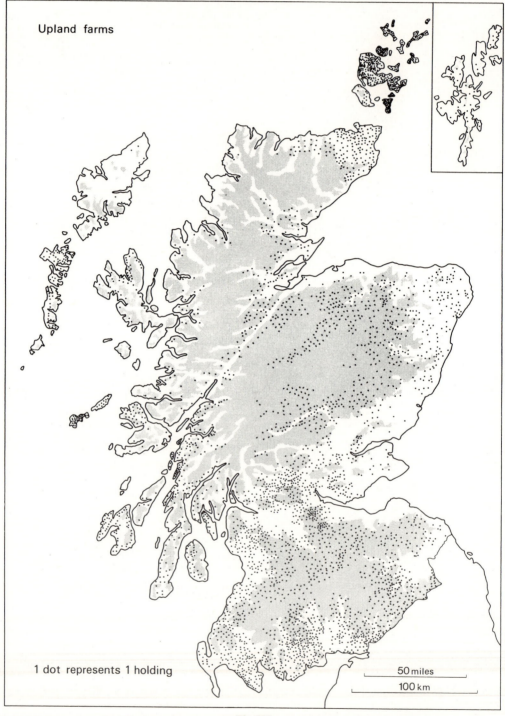

Upland farms

1 dot represents 1 holding

50 miles

100 km

Fig 207

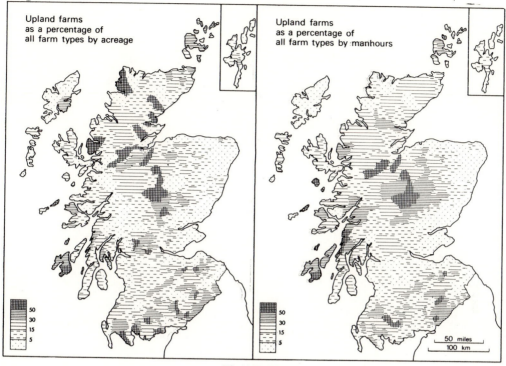

Figs 208–209

Upland Farms

Number: 3,771 in 1962; 5,060 in 1970

Percentage of full-time farms: 13·4 in 1962; 22·9 in 1970

Percentage of total area: 18·7 in 1962; 26·3 in 1970

Percentage of total smd: 9·6 in 1962; 15·5 in 1970

Upland farms are generally smaller than hill sheep farms and contain a greater proportion of improved land (15 per cent in 1962 and 22 per cent in 1970). Such farms were very widely distributed in 1962, but were especially characteristic of the upland margins, particularly in the eastern Highlands and in the Southern Uplands; they were also very numerous in Orkney (373 farms in 1962) and, to a lesser extent, in the plain of Caithness (Fig 207). Table 65 records the relative importance of such farms in each of the five administrative regions, expressed as a proportion of the Scottish total for this kind of farm, and again demonstrates the importance of North East and South West Scotland. As has been noted, the difference between proportions in 1962 and 1970 is partly due to a reclassifica-

TABLE 65

Regional Distribution of Upland Farms in 1962 and 1970

	High-land	North East	East Central	South East	South West	Scot-land
			Percentage by numbers			
1962	22	23	11	9	34	100
1970	14	31	10	10	36	100
			Percentage by total area			
1962	43	11	14	10	22	100
1970	32	23	13	10	22	100
			Percentage by smd			
1962	18	21	12	14	35	100
1970	12	27	10	15	36	100

Source: *Scottish Agricultural Economics*, Vol 15, and *Agricultural Statistics 1970*

tion of rearing with arable farms.

Figs 208 and 209 provide a more detailed analysis of regional differences, though the two maps are very similar. They show that in relative terms, upland farms were most important on the upland margins, even as shown by the large, often linear parishes; on both maps, the eastern Highlands and the Southern Uplands are particularly prominent.

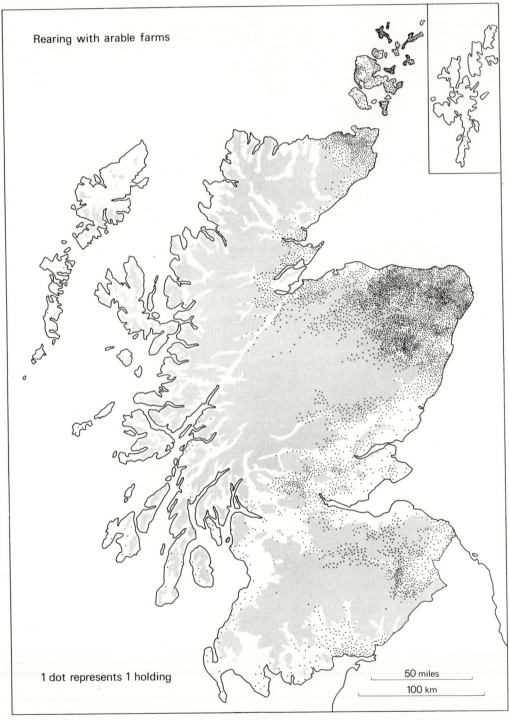

Rearing with arable farms

1 dot represents 1 holding

50 miles
100 km

Fig 210

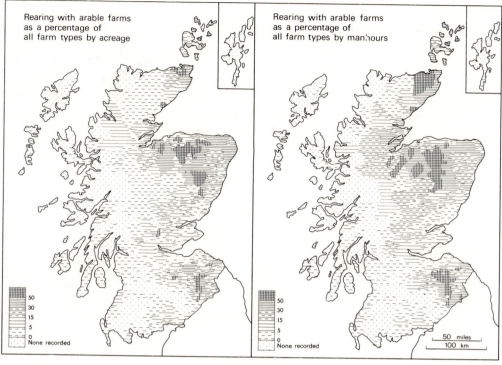

Figs 211–212

REARING WITH ARABLE FARMS

Number: 5,708 in 1962; 3,346 in 1970

Percentage of full-time farms: 20·2 in 1962; 15·2 in 1970

Percentage of total area: 9·9 in 1962; 5·8 in 1970

Percentage of total smd: 14·5 in 1962; 10·1 in 1970

The number of rearing with arable farms, which are smaller and have more improved land and more land under tillage, has also been affected by reclassification in 1970. Unlike the two preceding categories, but like other kinds of mixed farms, farms of this type have a markedly easterly distribution (Fig 210). The most prominent area was the north-east, especially Buchan, and most of the remaining farms were located along the upper margin of cultivation in the Tweed valley and the Forth and Tay basins, and in Orkney and the plain of Caithness; only in Dumfriesshire were they numerous in the south-west. Table 66 provides a measure of their relative importance in each of the five administrative regions, as measured by numbers, area and smd, and demonstrates the importance of

TABLE 66

Regional Distribution of Rearing with Arable Farms in 1962 and 1970

	High-land	North East	East Central	South East	South West	Scot-land
	Percentage by numbers					
1962	9	65	10	9	7	100
1970	7	66	9	10	9	100
	Percentage by total area					
1962	14	47	15	18	6	100
1970	11	45	15	20	9	100
	Percentage by total smd					
1962	8	56	13	17	7	100
1970	7	55	11	18	10	100

Source: *Scottish Agricultural Economics*, Vol 15, and *Agricultural Statistics 1970*

the North East Region by all three criteria.

Figs 211 and 212 (there is little difference between the two maps) confirm this easterly distribution and show that, both as a proportion of acreage and as a proportion of smd, such farms are most important around the upland margins in north-east Scotland and the Tweed–Teviot valley and in Caithness and Orkney.

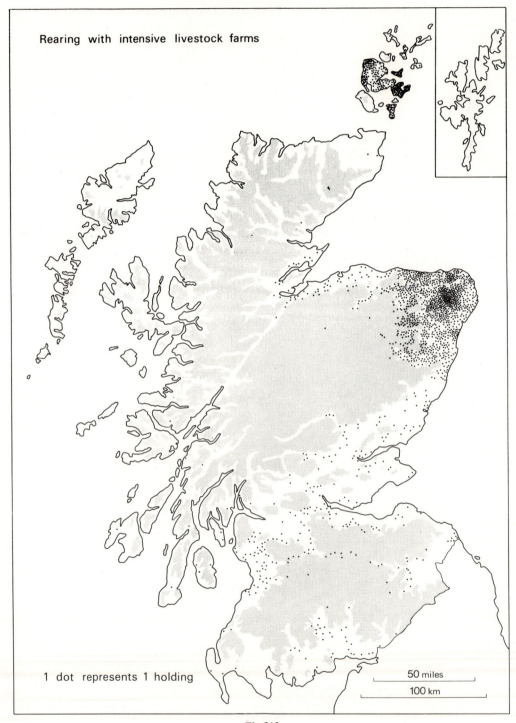

Rearing with intensive livestock farms

1 dot represents 1 holding

50 miles

100 km

Fig 213

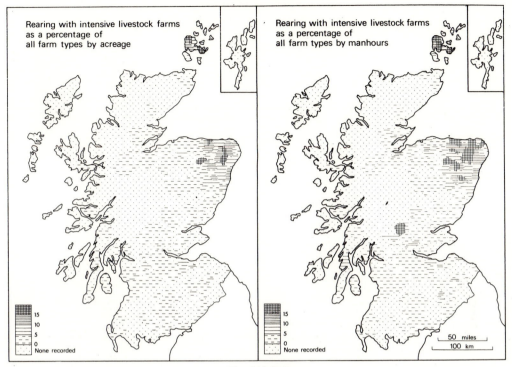

Figs 214–215

REARING WITH INTENSIVE LIVESTOCK FARMS

Number: 1,710 in 1962; 603 in 1970

Percentage of full-time farms: 6·1 in 1962; 2·7 in 1970

Percentage of total area: 0·9 in 1962; 0·9 in 1970

Percentage of total smd: 3·2 in 1962; 2·2 in 1970

The distinctive feature of this type of farm is the emphasis on pigs and poultry. Farms in this category form one of the smallest groups and their numbers have declined sharply since 1962. In that year they were concentrated very largely in two areas, north-east Scotland, especially Buchan, and Orkney, with small numbers scattered elsewhere throughout the lowlands (Fig 213); in 1962, the counties of Aberdeen and Banff had 61 per cent of all such farms and Orkney a further 15 per cent. Table 67 records the relative importance of this type in the five regions and shows the overwhelming dominance of the North East Region (which includes these three counties) on all three criteria. By 1970, however, some of this pre-eminence had been lost, though the North East remained by far the most important region for farms of this type.

TABLE 67

Regional Distribution of Rearing with Intensive Livestock Farms in 1962 and 1970

	High- land	North East	East Central	South East	South West	Scot- land
			Percentage by numbers			
1962	2	82	5	3	9	100
1970	4	67	9	7	13	100
			Percentage by total area			
1962	5	70	6	9	10	100
1970	25	43	9	12	12	100
			Percentage by total smd			
1962	3	78	7	5	7	100
1970	4	60	12	13	12	100

Source: *Scottish Agricultural Economics*, Vol 15, and *Agricultural Statistics 1970*

The importance of north-east Scotland (especially Buchan) and of Orkney is again confirmed by Figs 214 and 215, which are very similar, though only in one parish was rearing with intensive livestock the dominant type and in few did the proportion of either the acreage or man-days exceed 15 per cent. Over most of the uplands, no farms of this kind were recorded.

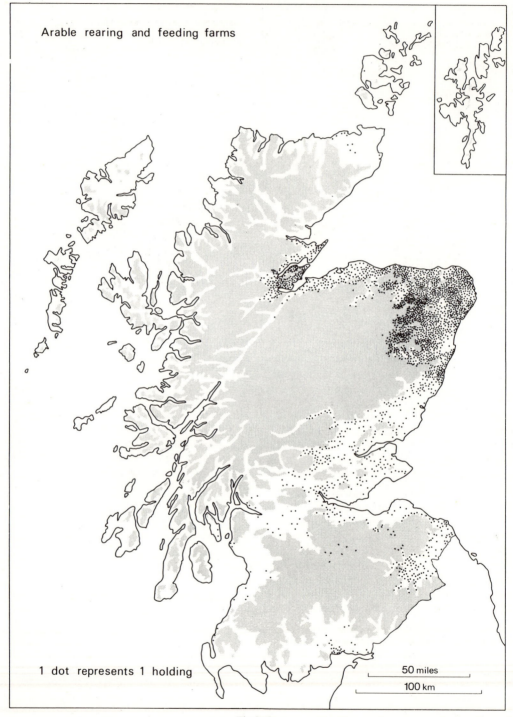

Arable rearing and feeding farms

1 dot represents 1 holding

50 miles

100 km

Fig 216

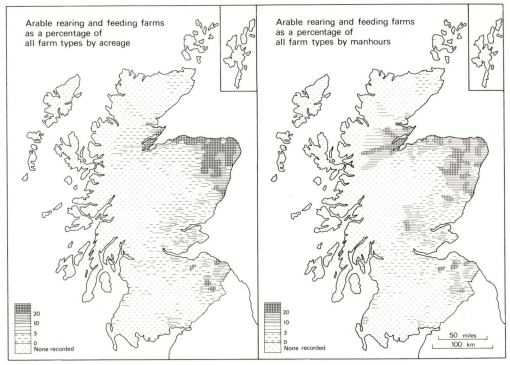

Arable rearing and feeding farms
as a percentage of
all farm types by acreage

Arable rearing and feeding farms
as a percentage of
all farm types by manhours

20
10
5
0
None recorded

20
10
5
0
None recorded

50 miles
100 km

Figs 217–218

ARABLE REARING AND FEEDING FARMS

Number: 2,356 in 1962; 1,215 in 1970
Percentage of full-time farms: 8·4 in 1962; 5·5 in 1970
Percentage of total area: 2·5 in 1962; 1·8 in 1970
Percentage of total smd: 6·2 in 1962; 4·0 in 1970

This is another category of mixed farm, but with a greater emphasis on crops for sale and the fattening of stock (see Table 62), and this type also underwent a sharp reduction in numbers in 1970. In 1962, such farms were most numerous in north-east Scotland, with 47 per cent in the county of Aberdeen alone; most of the remainder were in the lowlands of east-central Scotland and in the Merse, though this is the only type of mixed farm to be unimportant in Orkney, a reflection of the lesser suitability of these northern parishes for the production of sale crops (Fig 216). Table 68 records the relative importance of this type by number, area and man-days, and again shows the pre-eminence of the North East Region on all three counts, though the smaller average size of farms in this region is apparent. With the decline in numbers of farms

TABLE 68

Arable Rearing and Feeding Farms in 1962 and 1970

	High-land	North East	East Central	South East	South West	Scot-land
	Percentage by numbers					
1962	8	70	10	8	4	100
1970	7	66	12	10	5	100
	Percentage by total area					
1962	10	56	14	16	4	100
1970	8	52	16	19	5	100
	Percentage by total smd					
1962	9	59	16	12	4	100
1970	8	53	17	17	5	100

Source: *Scottish Agricultural Economics*, Vol 15, and *Agricultural Statistics 1970*

of this type there has been some reduction in this degree of concentration, though the North East remains by far the most important region.

Figs 217 and 218 show in more detail the relative importance of farms of this type as measured by area and smd. No farms in this category were recorded in a large number of upland parishes and the prominence of the north-east is again confirmed on both maps.

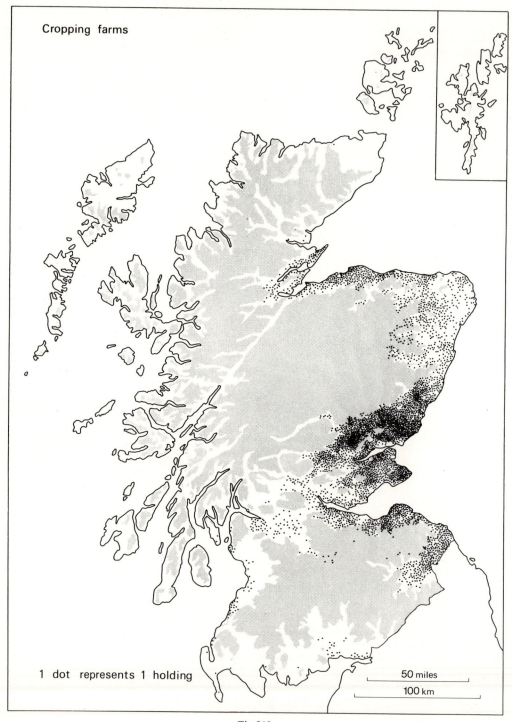

Cropping farms

1 dot represents 1 holding

50 miles

100 km

Fig 219

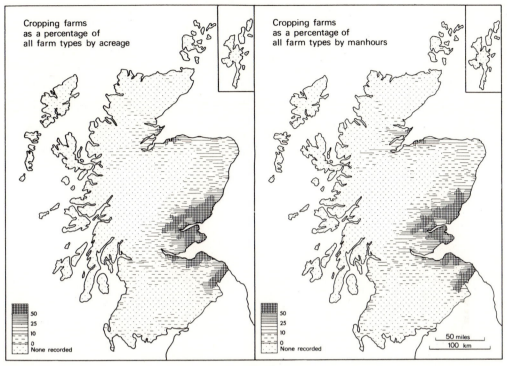

Figs 220–221

CROPPING FARMS

Number: 3,940 in 1962; 3,722 in 1970

Percentage of full-time farms: 14·0 in 1962; 16·9 in 1970

Percentage of total area: 5·7 in 1962; 7·1 in 1970

Percentage of total smd: 17·8 in 1962; 20·0 in 1970

More than half the gross output on cropping farms is derived from crops, with sales of fat cattle and sheep and of pigs and poultry as important ancillary sources (Table 62). It is not surprising, therefore, that such farms are largely concentrated in the lowlands along the east coast from Stonehaven to Dunbar (Fig 219); most of the remainder in 1962 were in the Merse, around the Moray Firth and in north-east Scotland. Table 69 records the proportion of cropping farms in the five regions; it should be noted that the North East Region includes Kincardine. The table shows the importance of the East Central Region and the more intensive nature of its farming; other notable features are the smaller extent of farms in the North East and the larger average area of those in the South East.

TABLE 69

Regional Distribution of Cropping Farms in 1962 and 1970

	High-land	North East	East Central	South East	South West	Scot-land
	Percentage by numbers					
1962	3	25	40	17	5	100
1970	4	26	48	17	5	100
	Percentage by total area					
1962	3	19	49	25	4	100
1970	5	21	47	24	4	100
	Percentage by total smd					
1962	3	18	54	22	4	100
1970	4	19	54	19	4	100

Source: *Scottish Agricultural Economics*, Vol 15, and *Agricultural Statistics 1970*

The pre-eminence of the eastern fringe of the central lowlands and the Merse is also revealed by Figs 220 and 221. As measured both by area and man-days, cropping farms predominated in the lowlands between Perth and Stonehaven, in eastern Fife, in East Lothian and in the Merse, as well as in small areas around the Moray Firth.

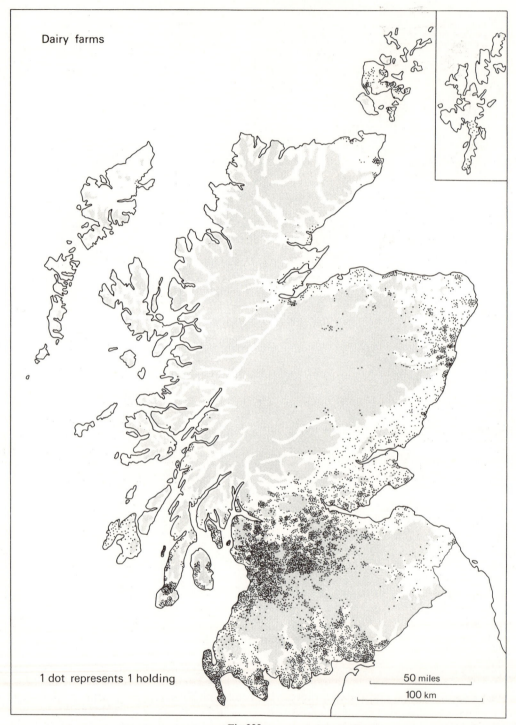

Dairy farms

1 dot represents 1 holding

50 miles

100 km

Fig 222

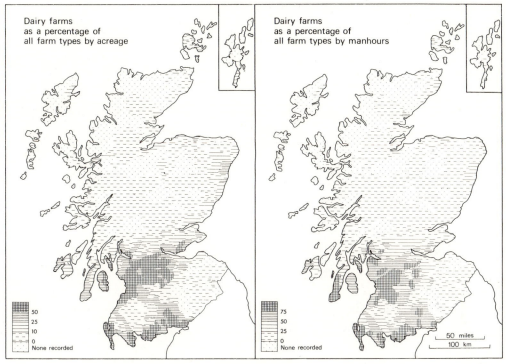

Figs 223–224

DAIRY FARMS

Number: 7,393 in 1962; 5,251 in 1970

Percentage of full-time farms: 26·2 in 1962; 23·8 in 1970

Percentage of total area: 10·8 in 1962; 9·0 in 1970

Percentage of total smd: 32·0 in 1962; 27·2 in 1970

Dairy farms were the most numerous category and one of the most intensive, though there was some decline, both in numbers and in relative importance, between 1962 and 1970. Dairy farms were associated most strongly with the western half of the central lowlands and the lowlands of south-west Scotland, though there were dairy farms scattered elsewhere throughout the lowlands and local concentrations, around Aberdeen (Fig 222). There were few such farms in south-east Scotland, in the Highlands or the islands, except in Arran and, to a lesser extent, Islay. Table 70 records the relative importance of dairy farms in the five regions and shows that about two-thirds of the farms of this type were in the South West Region in both years and on whichever basis of comparison they are considered; as with other types, dairy farms in the

TABLE 70

Regional Distribution of Dairy Farms in 1962 and 1970

	High-land	North East	East Central	South East	South West	Scot-land
	Percentage by numbers					
1962	6	11	9	6	68	100
1970	6	11	8	6	70	100
	Percentage by total area					
1962	9	9	8	6	67	100
1970	13	11	9	5	63	100
	Percentage by total smd					
1962	4	12	11	7	66	100
1970	5	13	11	5	66	100

Source: *Scottish Agricultural Economics*, Vol 15, and *Agricultural Statistics 1970*

Highlands tend to be both larger and less intensive.

The importance of the western half of the central lowlands and the lowlands of the south-west is again made clear by Figs 223 and 224 which show the relative contribution of dairy farms on a parish basis. Elsewhere dairying was important in relation to other types only around Aberdeen and in west Fife. There is little difference in the two distributions.

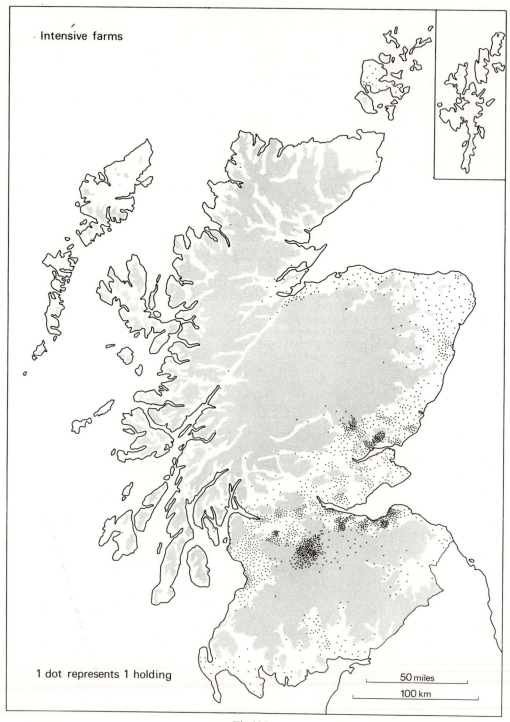

Intensive farms

1 dot represents 1 holding

50 miles

100 km

Fig 225

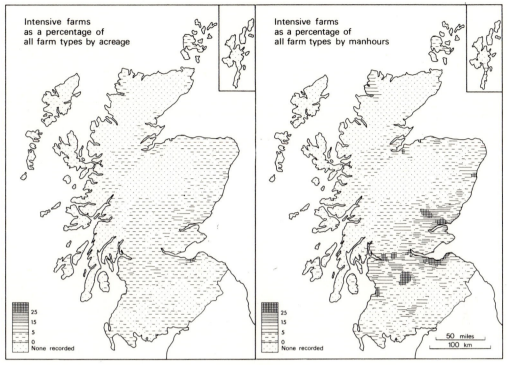

Figs 226–227

INTENSIVE FARMS

Number: 1,899 in 1962; 1,084 in 1970

Percentage of full-time farms: 6·7 in 1962; 7·2 in 1970

Percentage of total area: 0·4 in 1962; 0·5 in 1970

Percentage of total smd: 6·9 in 1962; 10·5 in 1970

These farms are mainly small in area but have the highest output per acre of all types. The term intensive also includes a wide variety of types and, of those identified in 1962, 49 per cent were horticultural farms (accounting for 60 per cent by smd), 18 per cent poultry farms, 18 per cent pig farms and 15 per cent mixed intensive farms. As Fig 225 shows, small numbers were scattered throughout the lowlands, but there were three major concentrations, in the Clyde valley, in the Lothians east and west of Edinburgh, and between Dundee and Blairgowrie. Table 71 records the relative importance of such farms in the five regions, though the considerable degree of localisation of intensive farms makes this a rather crude measure; for example, more than half the horticultural farms were in the counties of Angus, Lanark and Perth.

TABLE 71

Regional Distribution of Intensive Farms in 1962 and 1970

	High-land	North East	East Central	South East	South West	Scot-land
	Percentage by numbers					
1962	2	11	29	18	40	100
1970	3	17	30	16	35	100
	Percentage by total area					
1962	3	28	30	21	17	100
1970	3	16	33	18	30	100
	Percentage by total smd					
1962	1	11	32	17	39	100
1970	2	17	44	14	23	100

Source: *Scottish Agricultural Economics*, Vol 15, and *Agricultural Statistics 1970*

When the importance of intensive farms is assessed by their share of farmland, only a small number of parishes had more than 5 per cent in such farms (Fig 226). By contrast, they contributed more than 10 per cent of the man-days on all types of farm throughout most of the central lowlands and a quarter or more in the three areas revealed in Fig 227. Such farms were surprisingly widespread.

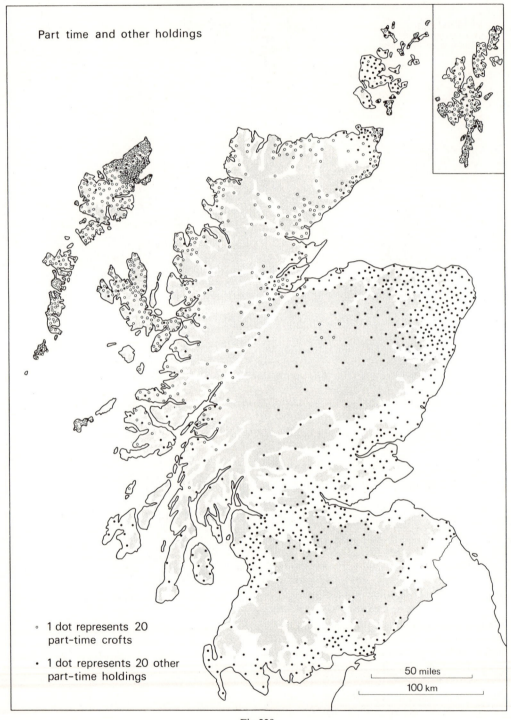

Part time and other holdings

∘ 1 dot represents 20
part-time crofts

• 1 dot represents 20 other
part-time holdings

50 miles

100 km

Fig 228

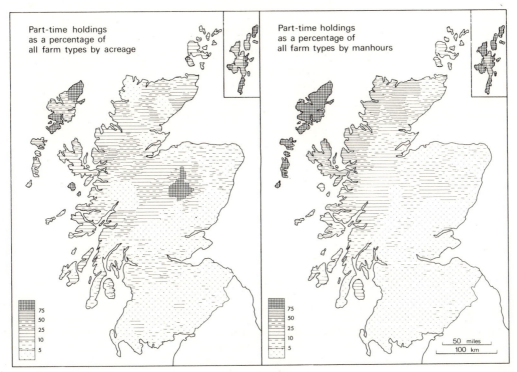

Figs 229–230

PART- AND SPARE-TIME HOLDINGS

Number of part-time: 11,848 in 1962; 5,895 in 1970

Number of spare-time: 20,904 in 1962; 10,292 in 1970

Percentage of all holdings: part-time 19·4 in 1962; 15·4 in 1970; spare-time 34·3 in 1962; 26·9 in 1970

Percentage of total area: part-time 5·5 in 1962; 6·6 in 1970; spare-time 8·3 in 1962; 4·7 in 1970

Percentage of total smd: part-time 4·4 in 1962; 4·1 in 1970; spare-time 1·7 in 1962; 2·2 in 1970

Holdings with fewer than 250smd are not classified by type, though they are understandably the most numerous class, even after the very considerable changes in 1970 which reduced their number to almost half that in 1962. Such holdings were widespread, but were particularly numerous in the islands and the north-west

mainland, where most were crofts (for only about 3 per cent of crofts had 250smds), in the north-east and in the central lowlands (Fig 228). Table 72 shows their relative contribution in the five regions and confirms the importance of the Highlands and the North East Region, a dominance which has been relatively little affected by the administrative changes in the census.

In relative terms, this dominance of the north-west is even more marked when the parish data are mapped. In most parishes in the Highlands and islands at least 10 per cent of the acreage in farms was in part- or spare-time holdings, a proportion which reached 50 per cent or more in the Outer Hebrides and Shetland (Fig 229). The importance of such holdings in these island groups was even more marked on the basis of their contribution to total smd which exceeded 75 per cent; in most of the western mainland such holdings also accounted for half or more (Fig 230).

TABLE 72

Regional Distribution of Part- and Spare-time Holdings in 1962 and 1970

		Highland	North East	East Central	South East	South West	Scotland
		Percentage by numbers					
1962	Part-time	49	28	7	4	12	100
	Spare-time	59	18	6	4	13	100
1970	Part-time	29	39	9	6	17	100
	Spare-time	60	21	5	4	11	100
		Percentage by total area					
1962	Part-time	68	19	5	2	7	100
	Spare-time	59	26	7	1	6	100
1970	Part-time	63	17	6	4	10	100
	Spare-time	76	13	2	2	7	100
		Percentage by total smd					
1962	Part-time	44	31	7	5	13	100
	Spare-time	53	19	8	4	16	100
1970	Part-time	27	41	9	6	17	100
	Spare-time	58	22	5	4	11	100

Source: *Scottish Agricultural Economics*, Vol 15, and *Agricultural Statistics 1970*

LEADING TYPE OF FARM

Although a very simplified map could easily be constructed, with dairying dominant in the western and south-western lowlands, cropping in the east, livestock rearing and feeding in the north-east and hill sheep in the uplands, no attempt has been made to integrate the information analysed in this section into a single type of farming map. Instead, a set of four maps is presented recording the leading or first ranking type of farm in each parish (or grouped parish). Figs 231 and 232 show, on the basis of acreage and man-days respectively, the leading type of holdings, including (where appropriate) part- and spare-time holdings. Figs 233 and 234, on the other hand, record the leading type in respect of full-time holdings only, and should be read in conjunction with Table 53 which shows similar information by regions. These maps, which should also be read in conjunction with the appropriate maps of individual types of farm, show those parishes in which the leading type accounted for 50 per cent or more. The regional pattern is broadly similar on all four maps, especially the dominance of cropping farms around the coasts of the Moray Firth and in the eastern part of the central lowlands, the Lothians and the Merse. On an acreage basis, hill sheep farms were the dominant type in both the Southern Uplands and the Highlands (though upland and rearing with arable farms were important in the eastern Highlands), while in the lowlands dairy farms were the leading type in the western half of the central lowlands and in the south-west, cropping farms the leading type in the east and south-east, and rearing with arable farms dominant in the north-east. Comparison on the basis of labour requirements greatly restricts the area over which hill sheep farms were the leading type, especially in the south-west where the area in which dairying is the leading type has been greatly extended; pockets of intensive farms appear as the first-ranking type around Aberdeen, Edinburgh and in the Clyde valley. On Fig 232, which shows leading types by smd, part- and spare-time holdings were sufficiently important to be the leading type in the Outer Hebrides, Shetland and Skye, as well as on the north-west mainland; by contrast, there were few areas where these holdings were sufficiently important to appear on the map of leading type by acreage (Fig 231). Apart from the exclusion of areas where part- and spare-time holdings were the leading type, there is little difference between the first and second pair of maps, though in all instances, rearing with arable farms will tend to have been replaced by upland farms in 1970.

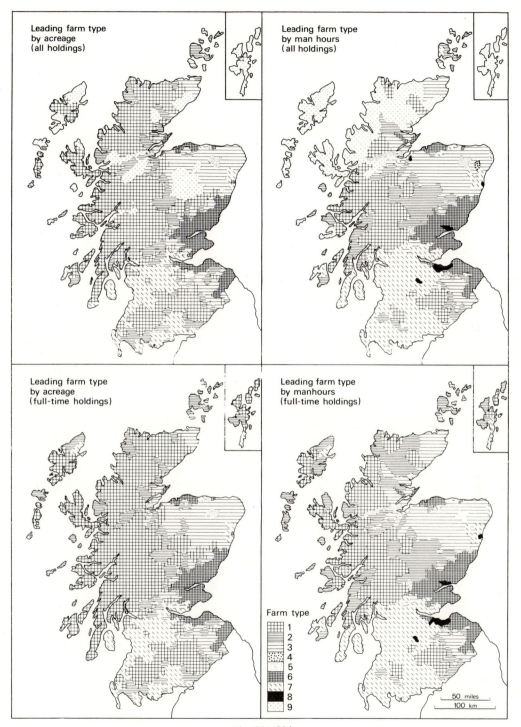

Leading farm type
by acreage
(all holdings)

Leading farm type
by man hours
(all holdings)

Leading farm type
by acreage
(full-time holdings)

Leading farm type
by manhours
(full-time holdings)

Farm type
1
2
3
4
5
6
7
8
9

50 miles
100 km

Figs 231–234

7

An Historical Perspective: 1870–1970

To conclude this atlas, it will be instructive to examine the antecedents of the 'present' agricultural geography (ie that in the mid-1960s, which has itself become part of the past during the preparation of this atlas), and to examine the stability or otherwise of the features which have been identified in the contemporary geography. This section does not attempt to present a definitive view, but merely to erect some signposts and indicate some of the main characteristics of the agricultural geography of Scotland at the beginning of the great agricultural depression in the last third of the nineteenth century and at the nadir of British farming in the 1930s. There are three main reasons for adopting such a limited objective. First, the agricultural returns themselves, particularly in 1870, were much more restricted in scope than they are today, especially in those branches of livestock husbandry which are of such importance in Scotland and on which even the contemporary returns can throw only a somewhat uncertain light. Secondly, the dominance of the uplands and rough grazings in shaping at least the spatial (if not the economic) characteristics of Scottish farming, together with the limited data available on upland farming, means that most of the changes which can be identified concern the lowlands and so cannot readily be portrayed by the method used in this atlas. Thirdly and most important, any attempt to paint a definitive picture of Scottish agriculture around 1870 and on the eve of World War II would require detailed investigations of primary sources.

THE AGRICULTURAL GEOGRAPHY OF SCOTLAND IN 1870

The agricultural census, as a continuing annual event, began in 1866, though the Highland and Agricultural Society had successfully collected statistics for all parts of Scotland from 1854 to 1857. The year 1870 was thus only the fifth year for which official statistics had been regularly collected, and there were bound to be inaccuracies, inconsistencies and omissions. It could not be claimed that the agricultural returns were as accurate in 1870 as they are today, for the prodecure of collection was more cumbersome and less familiar; there were many more holdings from which returns were sought (the lower limit then being a quarter of an acre) and there were many difficulties of interpretation to be resolved. Unfortunately, there are no independent national assessments of their reliability.

Despite the fact that only 0·5 per cent of Scottish farmers declined to make returns in 1866, there is no doubt that there were omissions in the early years of collection; the steady rise in the total acreage of crops and grass, from 4·16 million acres in 1866 (when the minimum holding was 5 acres) to 4·92 million acres in 1891, is itself fairly conclusive evidence; for while there were reports of reclamation in the hill areas, it seems highly unlikely that the acreage of improved land was in fact rising throughout the 1870s, the 1880s and the 1890s in the face of depressed conditions (admitted chiefly affecting arable farmers) and of the continuing loss of agricultural land to house building, railway construction, mining and the like. There is also reason to suppose that it was primarily the smaller holdings which tended to be omitted (and an occupier could hardly be expected to make a return unless he was asked for one) and that such holdings would often have a high proportion of permanent grass. There was also a good deal of confusion about the distinction between rough grazing, which was to be excluded from the returns, and permanent grass, which was to be included; for example, a clarification arising from a change of definition in 1889 led to a decrease in the acreage of permanent grass of some 2,200 acres (c 900ha) in

200

Haddington and of 2,400 acres (c 1,000ha) in Orkney. Given the nature of livestock farming in the Scottish uplands, it is also likely that an accurate statement of livestock numbers would have presented even greater difficulties to farmers than it does today. Acreages and numbers of livestock should therefore be treated with caution, though there is no reason to suppose that any major errors occurred or that the relative importance of the different enterprises (in so far as they can be identified) was affected.

More important for present purposes is the restricted range of items included in the 1870 census, when there were 34 items, compared with 125 in 1965; of the former 22 related to tillage crops and 9 to various categories of livestock. This lack of detail is particularly serious for the analysis of Scottish farming, especially the failure of the census to distinguish between beef and dairy cattle or between breeding ewes and other sheep one-year old and over which must then have included large numbers of wethers. No close analysis of dairying or beef breeding is possible on this basis and the treatment of sheep must be even more summary than it would have been in the late 1960s if the analysis in Section 4 had been based solely on information from the June census. Fortunately, the civil parish has been a much more stable unit than its counterpart in England and Wales, partly perhaps because of its greater size; as a result, it has been possible, with some manipulation of the parish summaries, to retain the 557 grouped parishes used in the analysis of the 1965 data for the investigations of the situation in both 1870 and 1938. The comparison is not exact, since there are uncertainties about data for some of the west Highland parishes; but any distortion is unlikely to be large.

The nature of the available data also affects the way in which the parish summaries of the 1870 agricultural census can be mapped. Because rough grazings were not included in the returns, it is not possible to calculate densities of livestock in relation to total agricultural land, so even greater reliance must be placed on dot maps. Reservations are also necessary about the propriety of computing arable land or permanent grassland as a percentage of the acreage under crops and grass because of uncertainties about the latter figure, but the subdivision of the

tillage acreage among the component crops is probably a reliable indicator which can safely be compared with the proportions today. The main reservations about the accuracy of the acreages returned as tillage relate to bare fallow, which has always presented problems of definition. Tillage is also a better indicator than arable land owing to uncertainties about the distinction between permanent and temporary grass; differences in agricultural practices in Scotland from those prevailing in most other parts of Great Britain have always presented difficulties in drawing a line between these two categories of grassland, and the distinction is likely to have been a somewhat arbitrary one.

The more generalised information given in the returns for 1870 also affects the comparability of individual items in the census with those in later censuses. Some (probably only a small part) of the acreage under the various cereals can be attributed to mixed corn, which was not separately recorded until 1918 and which was formerly attributed to its component crops. Many vegetables and several green fodder crops were also not separately enumerated and may be either included with other crops or omitted altogether, since no total acreage of tillage was recorded, the area under tillage being obtained only by adding up the acreages of the individual crops. Less detailed information on livestock has made it necessary to compute livestock units in a different way and with different components; thus horses were an important item in 1870 but were not recorded in 1965, while the reverse was true of poultry.

Nevertheless, while the picture of agriculture in the 1870s is biased towards crops, the essential features of Scottish agriculture broadly resemble those in 1965 (with the exceptions noted opposite the appropriate maps). The chief differences appear to be the greater importance of subsistence farming in the north-west and the lower densities of stocking (though this cannot be established from the maps). It should also be noted that these figures provide comparisons only of crop acreages and of livestock numbers and can take no account of the great differences in crop yields and livestock productivity which have taken place in the interval. It is in this respect that the most important changes have occurred.

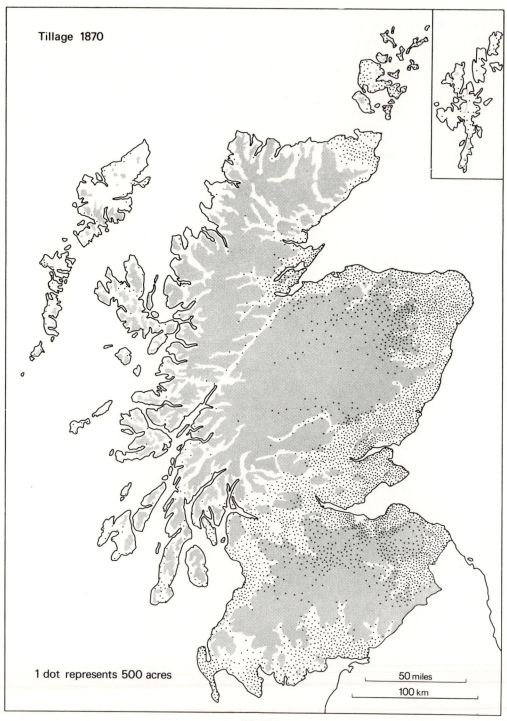

Tillage 1870

1 dot represents 500 acres

50 miles
100 km

Fig 235

CROPS AND GRASS

Area in 1870: 4,450,544 acres (1,801,110ha)

The distribution of crops and grass (ie improved land) in 1870 was broadly similar to that in 1965, although the acreage was 3 per cent higher in 1870. Although this difference is due in part to losses to urban and other related developments, there was undoubtedly some reversion of improved land to rough grazings, especially in the west and north; for example, the acreage in the Highlands fell by 9 per cent, or three times the national average in the same period. Unfortunately, interpretation is complicated by uncertainty about the extent to which the acreage under crops and grass was understated in 1870, owing to the failure to obtain returns from all holdings, though it is noteworthy that the acreage rose until 1891, when it stood at 4,917,380 acres (1,990,036ha), suggesting that a considerable part of the difference was due to better enumeration. There were also the ambiguities noted earlier about the division between rough grazing on the one hand and permanent and temporary grass on the other, and between the two kinds of grassland. In the circumstances, it seems unprofitable to attempt more detailed analyses of the total acreages of crops and grass, or of the different types of grassland. It is also impossible, since rough grazing was not then recorded, to examine changes in the distribution of total agricultural land. Attention will therefore be focused on crops other than grass.

TILLAGE

Area in 1870: 2,145,723 acres (868,362ha)

The acreage of tillage in 1870 was 41 per cent greater than that in 1965, a reflection in part of the loss of agricultural land to other uses, though the acreage in 1870 was almost certainly underestimated for the reasons already noted in the discussion of crops and grass. Although the main features of its distribution were broadly similar to those in the 1960s, it was more widely distributed (Fig 235); for while the North East, East Central and South East Regions had 78 per cent of the tillage acreage in 1965 (and 81 per cent in 1972), they had only 69 per cent in

TABLE 73

Crops and Grass and Tillage in 1870

High-land	North East	East Central	South East	South West	Scot-land
Percentage of tillage in each region					
9	30	22	17	21	100
Percentage of crops and grass in tillage in each region					
48	56	55	52	35	48

Source: Agricultural Returns 1870

1870 (Table 73). There was appreciably more cropped land in the islands and the lowlands of the south-west.

The distribution of tillage as a proportion of the acreage of crops and grass is recorded in Fig 240. While the map shows that at least 50 per cent of the improved land in the lowlands around the east coast was under tillage, some of the highest proportions were in crofting parishes in the north-west, and tillage accounted for half or more of the admittedly small acreages under crops in Orkney and in the Hebrides, compared with under 50 (and often 25) per cent in 1965 (Fig 53). Similarly, there were few upland parishes where less than 25 per cent of the improved land was under tillage compared with under a quarter in 1965.

Cropping was less diversified in 1870, with two crops, oats and turnips, accounting for 71 per cent of the land under tillage, and with wheat, barley and potatoes occupying a further 26 per cent. Of the remaining tillage bare fallow (of some importance in the west and north) accounted for 1 per cent and of the other crops grown (in descending order by acreage, vetches and other green crops; rye; cabbage; kohl rabi and rape; peas; flax; mangolds; and carrots) only beans occupied more than 1 per cent. Brassicas were noticeably less important than in 1965, while flax and rye are no longer returned separately; flax was grown in central, south and south-west Scotland, particularly in the counties of Lanark and Stirling. Orchards were recorded for the first time in 1871, at 6,865 (2,778ha), with nearly a third in Perthshire, though this figure is probably an understatement.

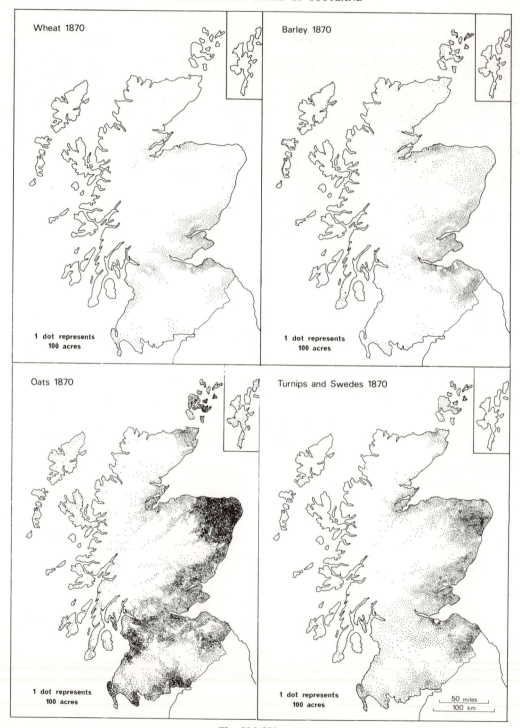

Figs 236–239

WHEAT

Area in 1870: 125,442 acres (50,766ha)

Wheat was absolutely, but not relatively, more important in 1870 than in the 1960s, accounting for 6 per cent of the land devoted to tillage in both years though the acreage had fallen by 28 per cent in 1965. The crop was more widely grown, particularly in the South West Region (which had only 6 per cent of the crop in 1965), notably on the coastal lowlands and around Glasgow (Fig 236); on the other hand, less wheat was grown in the North East Region than in 1965 (Tables 10 and 74). The Lothians, east Fife and the lowlands from Perth to Stonehaven were the most important area on both occasions.

BARLEY

Area in 1870: 244,142 acres (98,804ha)

Barley (including bere) was a relatively minor crop in 1870, accounting for only 11 per cent of the acreage under tillage that year, compared with 37 per cent in 1965 (when the acreage was more than twice as large). It was grown as far afield as Shetland and the Outer Hebrides where the crop was probably bere, a hardy four-rowed barley. These islands and the north-west Highlands apart, the main areas were the Lothians and the Merse, and to a lesser extent the eastern lowlands from the Forth to Stonehaven, with the East Central and South East Regions accounting for 32 and 29 per cent respectively (Fig 237). Barley was less important in the South West and in the North East Regions, which had 5 and 24 per cent respectively; for both wheat and barley were unpopular there as fodder crops, chiefly because of the quality of their straw.

OATS

Area in 1870: 1,019,596 acres (412,625ha)

Oats were by far the most important cereal; indeed, they were the most important crop grown in Scotland, accounting for 48 per cent of the acreage under tillage. The crop was widely grown throughout the lowlands and was especially important in Orkney and among the islands, though some was grown nearly everywhere (Fig 238); the two most important areas on the mainland were North East Scotland, which had 36 per cent of the acreage, and the South West (the Solway lowlands and Ayrshire), which had 26 per cent (Table 74). The widespread distribution and importance of the crop is indicated by the fact that in few areas was less than a quarter

of the tillage acreage under oats; in large parts of the uplands the proportions exceeded 50 per cent, though the actual acreages were small.

TURNIPS AND SWEDES

Area in 1870; 498,932 acres (201,915ha)

Turnips and swedes ranked as the second most important crop (by area) and occupied three times as large an acreage as in 1965, accounting for 23 per cent of the acreage under tillage. These crops were widely grown, but the North East Region was already pre-eminent (Fig 239), with 36 per cent of the national acreage; the eastern lowlands south of Stonehaven were also important.

POTATOES

Area in 1870: 180,169 acres (72,913ha)

Potatoes were less important in 1870 and are not shown by a dot map. They were widely grown, but the mapping of their relative importance is somewhat misleading in this connection, since it emphasises the role of the potato as a subsistence crop in the western Highlands and islands; indeed, the Highland Region accounted for 16 per cent of the acreage though it had only 9 per cent of that under tillage (Table 74). Nevertheless, the East Central Region was the most important region, as it was in the 1960s, though its pre-eminence was much less marked in 1870.

CROPPING

Table 74 summarises the regional contribution of the five main crops, showing the share of the acreage under each crop in each region.

TABLE 74

Tillage and the Principal Crops in 1870

High-land	North East	East Central	South East	South West	Scot-land
Percentage of tillage in each region					
9	30	22	17	21	100
Percentage of wheat in each region					
7	6	41	24	22	100
Percentage of barley in each region					
10	24	32	29	5	100
Percentage of oats in each region					
9	36	16	13	26	100
Percentage of turnips and swedes in each region					
7	36	20	20	17	100
Percentage of potatoes in each region					
16	13	31	14	25	100

Source: Agricultural Returns 1870

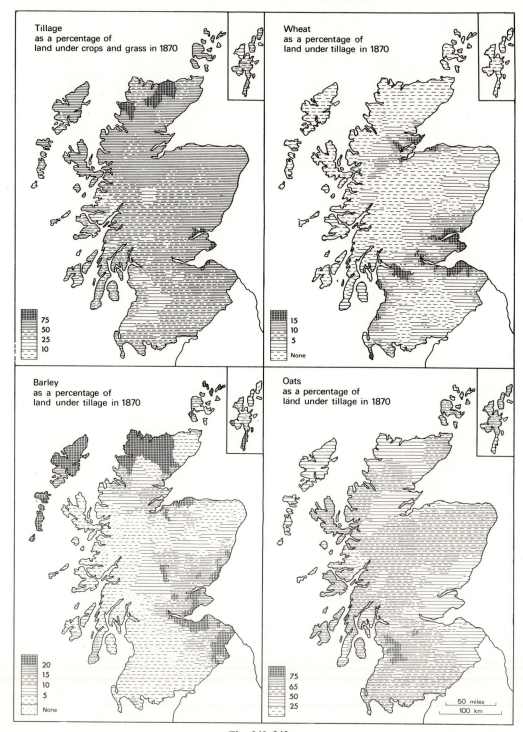

Figs 240–243

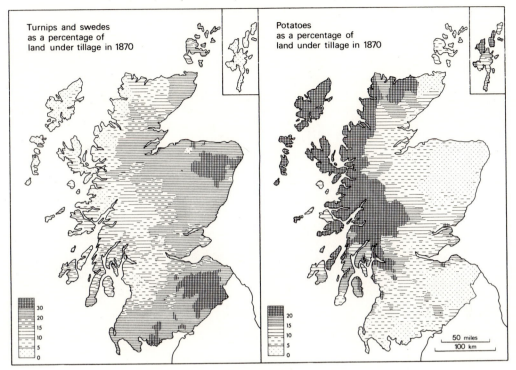

Figs 244–245

Table 75 shows the relative importance of each of the five main crops, expressed as a proportion of the acreage under tillage; to provide a reference point, the proportion of the crops and grass acreage under tillage is also given (though, for reasons already noted, this parameter should be treated with some caution).

The most striking feature of the map of tillage is the high proportion in western and Highland districts (though the acreages were generally small); in only a few areas was less than 25 per cent under tillage (Fig 240). Fig 241 shows that wheat was relatively more important in the Lothians and in Tayside, Fig 242 that barley had a dichotomous distribution, in the south-east and the north-west, while oats were relatively most important in the north-east and the south-west (Fig 243). Turnips and swedes and potatoes, though widely grown, show contrasting distributions, for turnips were relatively more important in the north-east and south-east (Fig 244), and

potatoes (chiefly in their subsistence role) in the north-west and in most of the islands (Fig 245).

TABLE 75

Tillage and the Principal Crops in 1870

	High-land	North East	East Central	South East	South West	Scot-land
Percentage of the crops and grass acreage under tillage						
	48	56	55	52	35	48
Percentage of tillage under wheat						
	4	1	11	8	6	6
Percentage of tillage under barley						
	12	9	17	19	3	11
Percentage of tillage under oats						
	47	56	35	35	58	48
Percentage of tillage under turnips and swedes						
	18	28	21	27	18	23
Percentage of tillage under potatoes						
	14	4	12	7	10	8

Source: Agricultural Returns 1870

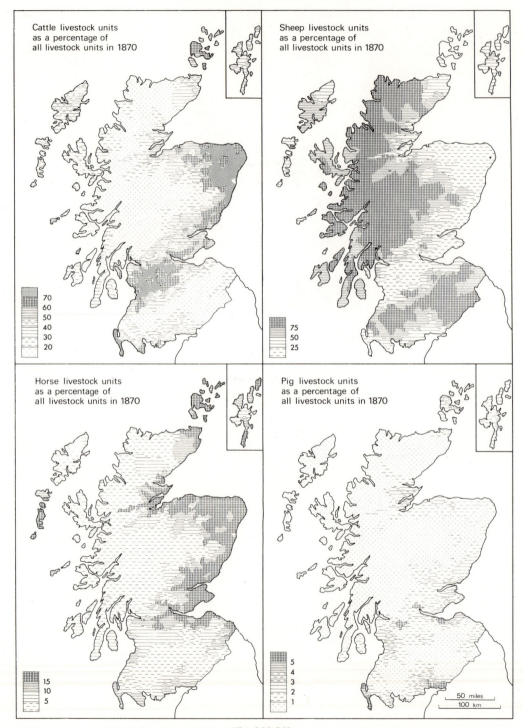

Cattle livestock units
as a percentage of
all livestock units in 1870

70
60
50
40
30
20

Sheep livestock units
as a percentage of
all livestock units in 1870

75
50
25

Horse livestock units
as a percentage of
all livestock units in 1870

15
10
5

Pig livestock units
as a percentage of
all livestock units in 1870

5
4
3
2
1

50 miles
100 km

Figs 246–249

LIVESTOCK

Number of livestock units in 1870: 1,970,517

The categories of livestock returned in 1870 were much more generalised than those recorded in 1965; only 3 classes of cattle were recognised, 2 classes of sheep and only the total number of pigs, compared with 21 classes of cattle, 5 of sheep and 8 of pigs. Poultry were not recorded in 1870 but there were 3 classes of horses (which were no longer enumerated in 1965). It was not therefore possible to use the factors employed in 1965 and 1972 for the computation of livestock units, and the following factors were used: cows and heifers in calf and milk, 1; other cattle two years old and over, 0.75; other cattle under two years old, 0.375; sheep one year old and over, 0·20; sheep under one year old, 0·067; pigs, 0·15; and horses, 1. Table 76 records the distribution of the different classes of livestock in 1870 (expressed as livestock units) by regions and their regional share of the total number of livestock units, while Figs 246–9 show the latter information by parishes.

CATTLE LIVESTOCK UNITS
Number in 1870: 720,987

Cattle, accounting for 37 per cent of all livestock units, were most important, as they are today, in the north-east and in the western half of the central lowlands, accounting for half or more of all livestock in both areas, though they were relatively less important in the coastal lowlands of the south-west (Fig 246). Unfortunately, it is not possible to show beef and dairy cattle separately.

SHEEP LIVESTOCK UNITS
Number in 1870: 720,987

The relative importance of sheep also showed a very similar pattern of distribution in 1870 and 1965, with sheep accounting for three-quarters or more of all livestock in the Southern Uplands and throughout the western highlands (Fig 247; cf Fig 170). Sheep were the leading livestock everywhere in the uplands and in most of the islands except the Orkneys; the lowest proportions were in the eastern arable areas, with the exception of the Merse.

TABLE 76
Livestock Units in 1870

	High-land	North East	East Central	South East	South West	Scot-land
	Numbers in thousands					
	544	308	288	267	564	1,971
	Percentage of livestock units in each region					
	28	15	15	14	29	100
	Percentage of cattle livestock units in each region					
	17	26	15	7	34	100
	Percentage of sheep livestock units in each region					
	37	6	13	18	25	100
	Percentage of horse livestock units in each region					
	16	29	19	12	23	100
	Percentage of pig livestock units in each region					
	13	17	17	13	41	100
	Percentage of each category in each region					
Cattle	23	62	38	19	44	37
Sheep	71	21	49	72	48	53
Horses	5	16	12	8	7	9
Pigs	1	1	1	1	2	1

Source: Agricultural Returns 1870

HORSE LIVESTOCK UNITS

Number in 1870: 172,871

Horses, as the main form of both traction and transport, were widespread, being most numerous and relatively most important in the North East (Fig 248), and generally in the eastern arable regions (though this is not so apparent from Table 76 owing to the heterogeneous nature and variable size of the regions). Horses were relatively least important in the uplands.

PIG LIVESTOCK UNITS

Number in 1870: 23,804

Pigs were of minor importance in all regions and accounted for 5 per cent or more of all livestock units only in a very few scattered parishes (Fig 249). There was some association with large cities and with the areas of dairy farming in the western half of the central lowlands, but no sign of the pre-eminence of the north-east, which is a marked feature of pig production in the 1960s.

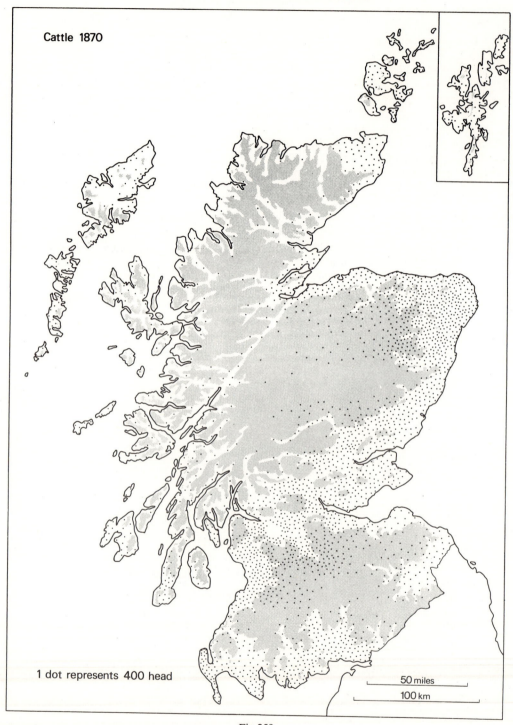

Cattle 1870

1 dot represents 400 head

50 miles
100 km

Fig 250

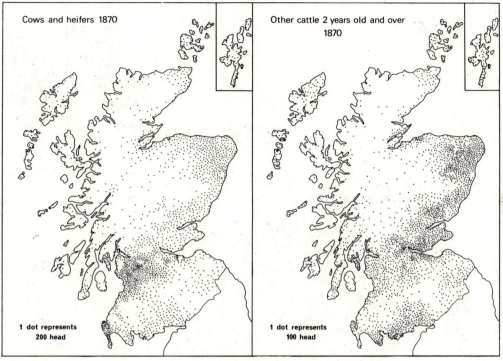

Figs 251–252

CATTLE

Number in 1870: 1,041,434

Fig 250 shows the distribution of cattle in 1870 and employs the same size of dot as in the corresponding map for 1965 (Fig 127), when cattle were twice as numerous. They were found in large numbers throughout the lowlands, being least important in the Merse and the Lothians, a fact confirmed by the low proportion of the national herd in the South East Region (Table 77). Cows and heifers may broadly be equated with dairy cows, though they include animals rearing calves for beef; in any case, the distinction between beef and dairy cattle was less clear cut then than it is today. Partly for this reason and partly because of the smaller size of the market for liquid milk, the pre-eminence of the South West was less marked than it is now (Fig 251); the main areas were the western half of the central lowlands and the Rhinns of Galloway.

While other cattle two years old and over may include dry cows and culled cows being fattened for slaughter, they can be used as an indicator of the importance of fattening, since most cattle

TABLE 77

Cattle in 1870

High-land	North East	East Central	South East	South West	Scot-land
Percentage of cattle in each region					
16	28	16	7	32	100
Percentage of cows and heifers in each region					
19	23	12	6	40	100
Percentage of other cattle two-years old and over in each region					
16	26	21	9	28	100
Percentage of other cattle under two-years old in each region					
14	34	17	7	28	100

Source: Agricultural Returns 1870

were over two years old at the time of slaughter (Fig 252). In 1870 the eastern regions had a large share of such cattle, though proportions were relatively low in the Merse and the Lothians (Table 77). It should be remembered, however, that this map portrays a summer distribution and that many cattle would have been fattened in yards in winter. For younger cattle, the North East, with its large number of small family farms, was pre-eminent.

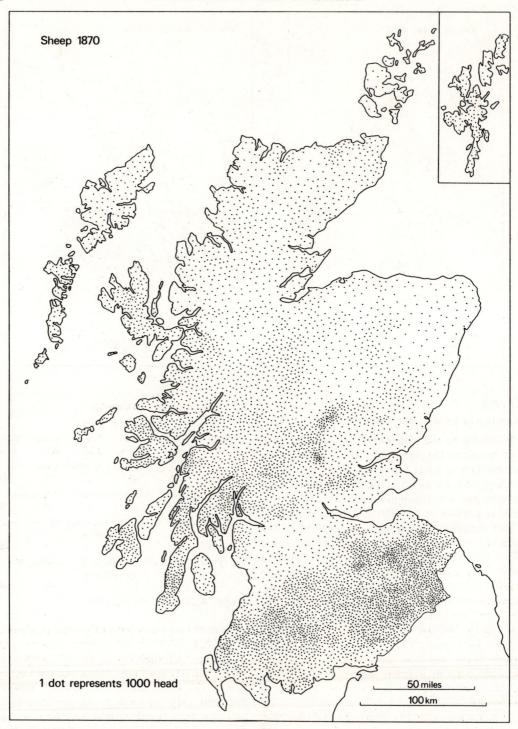

Sheep 1870

1 dot represents 1000 head

50 miles

100 km

Fig 253

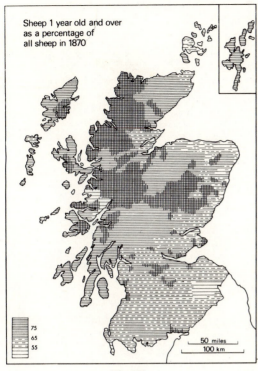

Sheep 1 year old and over
as a percentage of
all sheep in 1870

75
65
55

50 miles
100 km

Fig 254

TABLE 78

Sheep in 1870

High- land	North East	East Central	South East	South West	Scot- land
Percentage of sheep in each region					
35	6	13	20	26	100
Percentage of sheep one-year old and over in each region					
38	6	14	17	25	100
Percentage of sheep under one-year old in each region					
28	6	12	24	30	100

Source: Agricultural Returns 1870

The general pattern was broadly similar, with the Southern Uplands and the Merse as the principal areas; other notable features were the much smaller importance of sheep in Caithness and in north-east Scotland compared with the present, though sheep were generally less important in the lowlands. Table 78 confirms this distribution. Breeding ewes were not distinguished from other sheep one-year old and over in the agricultural census in 1870, a deficiency which makes interpretation difficult, since older wethers were then a much more important element in sheep flocks. Some indication of the greater importance of lambs in the more favoured areas of the south and east is, however, given by Fig 254, which shows the percentages of older sheep (or, if the key is read as the complement of 100, of younger sheep). The map reveals a broad contrast between the Highlands and other parts of Scotland.

SHEEP

Number in 1870: 6,750,854

Fig 253 shows the distribution of sheep in 1870 and employs the same size of dot as in 1965, when sheep were 27 per cent more numerous (though some allowance must be made for improvements in the accuracy of the returns).

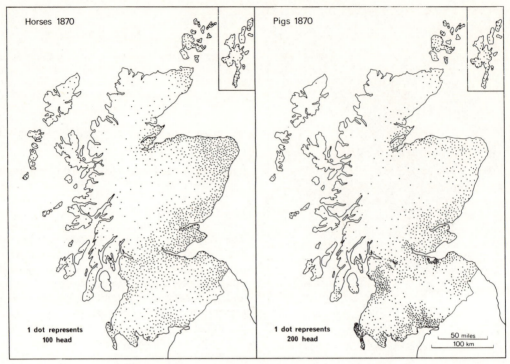

Figs 255–256

HORSES

Number of horses in 1870: 172,871

Fig 255 confirms the widespread distribution of horses throughout the lowlands; only in the uplands were numbers small. Table 79 shows the share of each region.

PIGS

Number of pigs in 1870: 158,690

The distribution of pigs was much more patchy, with the largest numbers in the lowlands of Ayrshire and the south-west, and with highly localised concentrations around Edinburgh and Glasgow (Fig 256). Table 80 confirms the regional importance of the South West, in marked contrast to the situation in 1972 when the North East was the most important region.

TABLE 79

Horses in 1870

High-land	North East	East Central	South East	South West	Scot-land
Percentage of horses in each region					
16	29	19	12	23	100

Source: Agricultural Returns 1870

TABLE 80

Pigs in 1870

High-land	North East	East Central	South East	South West	Scot-land
Percentage of pigs in each region					
13	17	17	13	41	100

Source: Agricultural Returns 1870

THE AGRICULTURAL GEOGRAPHY OF SCOTLAND IN 1938

The 1930s mark the end of a long period in which the agricultural industry, if not always depressed, was always disadvantaged in relation to other sectors of the British economy. Alone of the major European powers, the United Kingdom had followed a policy of laissez-faire during the preceding fifty years and had allowed virtually unrestricted access to imports of agricultural produce; and, while urban and industrial Britain benefited from a policy of cheap food, the agricultural community did not. The acreage under crops fell to the lowest figure it had ever reached since statistics were first recorded, but the statistics cannot reveal the extent to which fields and buildings had been neglected and farmers had lost heart. In the wake of the world economic crisis of 1929–31, some measures were taken to help farmers; they included powers which permitted the creation of producer marketing boards (and those formed to control the marketing of milk were especially important), and the introduction of guaranteed prices for wheat and fatstock, of quotas and tariffs on imports of horticultural produce, of subsidies on lime, and finally in 1939, in anticipation of the outbreak of hostilities, of grants to encourage the ploughing of older pasture. These measures had barely begun to take effect before the outbreak of World War II, which was followed by revolutionary changes, both technical and economic, in the character of British agriculture. The late 1930s can thus be regarded as a major divide in agricultural history.

Scottish agriculture shared these experiences, although the greater emphasis on livestock and the long-established use of long leys made changes in Scottish agriculture less necessary than they were in English agriculture; on the other hand, the range of opportunities was generally smaller and the limited extent of good land meant that Scottish farmers were not well placed to weather agricultural depression. It is true that Scotland is more nearly self-sufficient in temperate agricultural products and is a net exporter of store and fat sheep and of fat cattle and that this tendency was, to some extent, accentuated by the developments of the 1930s,

particularly by the creation of the three milk marketing boards; for, by agreement with the Milk Marketing Board (which has responsibility for sales of milk in England and Wales), Scotland became virtually self-contained for milk marketing and the considerable trade in liquid milk between south-west Scotland and north-east England was ended. On the other hand, improvements in transport were beginning to deprive Scottish horticulturalists of some of the advantages of proximity to markets which they had previously enjoyed over producers in England.

By June 1938 the agricultural census had largely achieved its present form. In place of the 34 items recorded in June 1870 there were 100, of which 47 related to livestock and 7 to labour. One omission had been repaired when the regular collection of statistics of poultry was begun in 1925, beef and dairy cattle were separately distinguished, and more meaningful categories of pigs and sheep had been adopted, although, as is so often the case, figures for the first few years following such changes must be treated with caution. The available data thus permit a fuller analysis of the agricultural scene than was possible for 1870, but such an analysis cannot be undertaken within the compass of this book and only an outline of the main features will be given here, with an emphasis on those items which can be compared with the 1870 and 1965 cross-sections. Of course, rigid adherence to this rule is not always possible, particularly in respect of the division into beef and dairy cattle.

Fortunately, this summary view can be supplemented by other published sources, two of which merit particular attention. In 1931, H. J. Wood produced his pioneer *Agricultural Atlas of Scotland*, which is based on agricultural census data for 1927 and thus portrays the agricultural geography of the late 1920s. Secondly, there are the maps and county reports of the Land Utilisation Survey, which provide coverage for the whole of Scotland on a fairly uniform basis and also make use of the material from the agricultural censuses (though the degree to which this was done varies with the author). Together, these various studies provide a baseline from which subsequent agricultural advances can be measured.

LAND UTILISATION IN 1938

AGRICULTURAL LAND

Area in 1938: 15,008,253 acres (6,073,757ha)

From 1892 onwards it is possible to form some estimate of the total extent of agricultural land, since records of the area of rough land used for grazing were first collected on a regular basis in that year. Unfortunately, the nature of these records is such that no safe conclusions can be drawn about changes over time, but they do serve as a reminder of the large part of Scotland that is unimproved agricultural land.

ROUGH GRAZING

Area in 1938: 10,447,873 acres (4,228,196ha)

No separate map of rough grazing has been included in this section, nor has the total area of agricultural land been used as the base for the calculation of stocking densities, largely because comparisons between 1965 and 1938 would not be possible owing to changes in the nature of what is included under the heading 'rough grazing'; for, whereas in 1965 the area returned as rough grazing included the total area of deer forests, whether they were used for grazing or were capable of being used for grazing or not, the acreage of rough grazing returned in 1938 included only those parts of deer forests which were grazed or were capable of being grazed, a figure which was perhaps $1\frac{1}{2}$ million acres (600,000ha) less. Of the acreage of deer forests in 1939, some 38 per cent was thought to be grazed, though only 29 per cent was on deer forests for which stock was returned. Fortunately, the changes in the distribution of rough grazings, however defined, have been relatively small, and have largely been confined to improvements and reversions around the fringes of the uplands, though about a million acres (c 400,000ha) has been taken for afforestation since 1919.

CROPS AND GRASS

Area in 1938: 4,560,380 acres (1,845,561ha)

The recorded area of improved land (crops and grass) has been fairly stable for most of the period for which statistics have been collected,

particularly before World War I, when losses to urban development were at least in part compensated by improvements in the coverage of the returns. Changes in definition again preclude any close analysis, though, at face value, the acreage of crops and grass in 1938 was 7 per cent greater than in 1965. Table 81 records the proportions of the area of improved land in each region in 1938, values which do not differ greatly from those recorded in 1965. Most of this land was under improved grassland of one kind or another, a fact also confirmed by Table 73, which records, for each region, the proportion of crops and grass under tillage (and hence by subtraction from 100, the proportion under grass). The rank order of the regions is the same as in 1870, but the proportions in Scotland and in each of the five regions are all lower, notably in the South East Region (cf Table 73).

TABLE 81

Crops and Grass and Tillage in 1938

High-land	North East	East Central	South East	South West	Scot-land
Percentage of crops and grass in each region					
10	27	18	16	30	100
Percentage of tillage in each region					
9	34	23	15	18	100
Percentage of crops and grass in tillage					
32	43	43	33	20	34

Source: *Agricultural Statistics 1938*

TILLAGE AND PRINCIPAL CROPS

Area under tillage in 1938: 1,523,939 acres (616,730ha)

Area under wheat in 1938: 92,497 acres (27,433ha)

Area under barley in 1938: 98,928 acres (40,036ha)

Area under oats in 1938: 797,752 acres (322,846ha)

Area under potatoes in 1938: 134,758 acres (54,536ha)

Area under turnips and swedes for stockfeeding in 1938: 320,241 acres (129,600ha)

As Table 81 has shown, land used for tillage was to be found mainly in the East Central and North East Regions. As in 1870, cropping was

TABLE 82

Tillage and Principal Crops in 1938

	Highland	North East	East Central	South East	South West	Scotland
Percentage of the acreage under tillage						
Wheat	1	1	13	12	4	6
Barley	6	7	5	15	—	6
Oats	55	60	43	35	62	52
Potatoes	9	3	16	9	10	9
Turnips and swedes	18	26	17	22	18	21
All crops	100	100	100	100	100	100

Source: *Agricultural Statistics 1938*

dominated by oats, turnips and swedes (though the latter were not then subdivided into those for human consumption and those for stock-feeding), which together accounted for 73 per cent of the land under tillage, with wheat, barley and potatoes occupying a further 21 per cent. Although horticultural crops were recorded in 1938, they accounted for only 14,457 acres (5,851ha), or less than 1 per cent of the tillage acreage, with soft fruit occupying 70 per cent of this area; land under orchards had fallen to 1,376 acres (557ha), a third of which were in Lanarkshire. Sugar beet, with 7,391 acres (2,891ha), was the only other crop of any importance. The remaining crops were nearly all fodder crops, in descending order of importance (as measured by acreage): cabbage and kale, rape, vetches and other green crops, mangolds, beans, mixed corn and rye. Bare fallow accounted for 18,265 acres (7,392ha), or just under 1 per cent. This pattern of cropping much more strongly resembled that in 1870 than cropping in 1965, when oats and turnips were losing their pre-eminence and barley had become the principal cereal and the leading crop.

As in 1870, there were none the less regional differences in cropping and the broad contrasts in the five regions are shown in Table 82. Allowance must, of course, be made for the heterogeneity of these regions, and more detailed comparisons are possible from an examination of Figs 261–5. The table reveals the much greater importance of wheat in the East Central and South East Regions, and the relative importance of barley in the South East and of potatoes in the East Central Region. Oats and turnips were the leading crops in each region, though their relative importance varied.

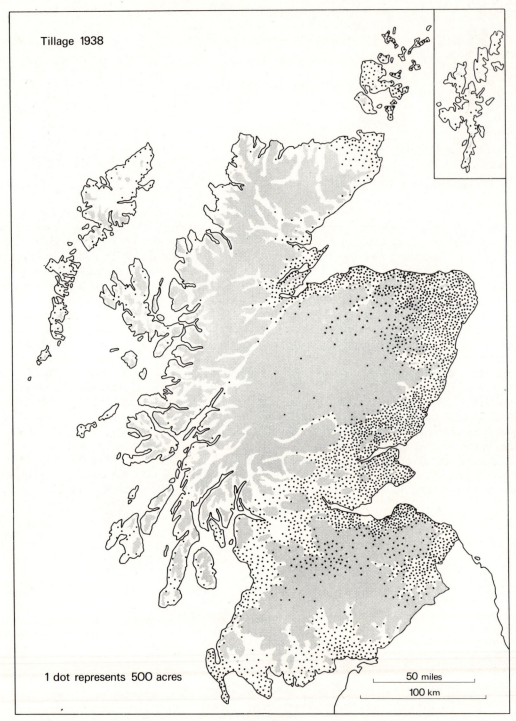

Tillage 1938

1 dot represents 500 acres

50 miles

100 km

Fig 257

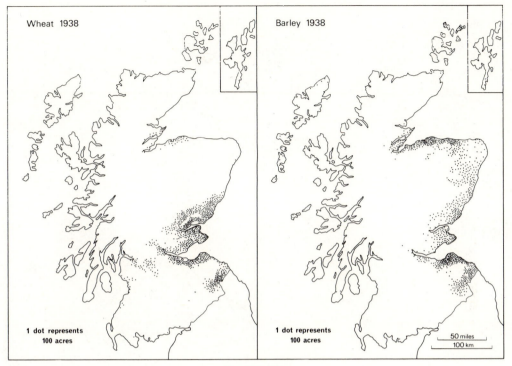

Figs 258–259

TILLAGE AND PRINCIPAL CROPS

Figs 257, 258 and 259 show the distribution of land under tillage, wheat and barley respectively in 1938. The tillage acreage was only 71 per cent of that in 1870 and was to be found mainly in eastern Scotland; but its distribution was surprisingly widespread, with considerable acreages grown in the lowlands of south-west Scotland and in the islands. The wheat acreage, which was 73 per cent of that in 1870, was largely confined to the lowlands around the Forth and Tay estuaries and to the Merse; barley, of which the acreage had declined to only 41 per cent of the 1870 total, was more widely grown in the eastern lowlands, with relatively large acreages around the Moray Firth. Table 83 records the regional distribution of the principal crops, with the percentage of tillage providing a reference point; wheat and potatoes were clearly the most localised crops.

TABLE 83

Tillage and Principal Crops in 1938

High-land	North East	East Central	South East	South West	Scot-land
Percentage of tillage in each region					
9	34	23	15	18	100
Percentage of wheat in each region					
2	5	50	31	12	100
Percentage of barley in each region					
9	39	18	34	1	100
Percentage of oats in each region					
10	40	19	10	21	100
Percentage of potatoes in each region					
10	13	42	15	20	100
Percentage of turnips and swedes for stockfeeding in each region					
8	43	19	15	15	100

Source: *Agricultural Statistics 1938*

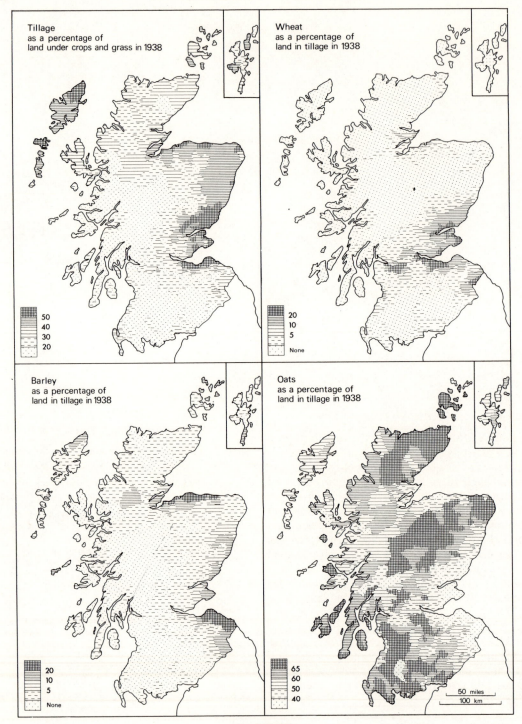

Tillage
as a percentage of
land under crops and grass in 1938

50
40
30
20

Wheat
as a percentage of
land in tillage in 1938

20
10
5
None

Barley
as a percentage of
land in tillage in 1938

20
10
5
None

Oats
as a percentage of
land in tillage in 1938

65
60
50
40

50 miles
100 km

Figs 260–263

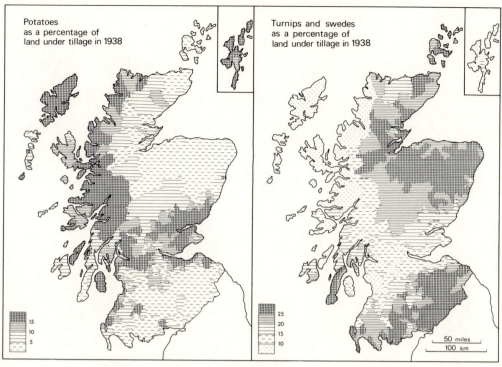

Figs 264–265

TILLAGE AND PRINCIPAL CROPS

Figs 260–5 provide a comparative view of the distribution of tillage and principal crops (see also Table 82). The proportion of the crops and grass acreage under tillage exceeded 50 per cent in two contrasting areas, the lowlands of the East Central Region and the islands, though the acreages there were small (Fig 260); the Merse was distinctive among the eastern lowlands, with only 30–40 per cent under tillage, whereas the proportion exceeded 40 per cent throughout the north-east, as it also did in the Outer Hebrides. Wheat accounted for 20 per cent or more of the tillage acreage in a small number of parishes around the Forth and Tay estuaries and in the Clyde valley; in few areas outside the eastern lowlands did the proportion exceed 10 per cent (Fig 261). As Fig 262 shows, barley was also prominent in two regions, the Merse and East Lothian, and the lowlands around the Moray Firth; over large parts of Scotland no wheat or barley was grown. By contrast, in only a few areas (mainly those where wheat was important) was less than 40 per cent of the tillage acreage under oats. The north-east and south-west (including the islands) were the principal areas; Orkney was again distinctive among the northern islands, with a high proportion of land under oats (Fig 263). Potatoes were relatively important in two areas, the Forth–Tay lowlands of the East Central Region, where they were primarily a cash crop for both seed and ware, and the western Highlands and some of the islands, where they were mainly a subsistence crop; those parishes on the coast of Ayrshire where early potatoes were grown also stand out (Fig 264). Turnips and swedes for stockfeeding, on the other hand, were most characteristic of the north-east and south-east, where the feeding of sheep and cattle was important (Fig 265).

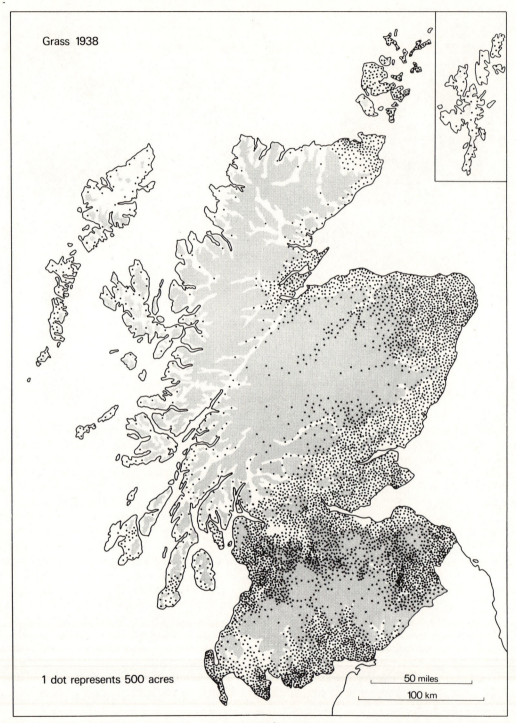

Grass 1938

1 dot represents 500 acres

50 miles

100 km

Fig 266

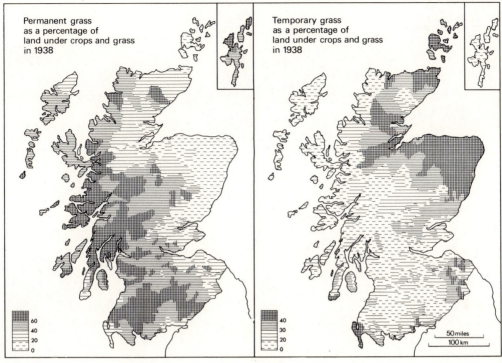

Figs 267–268

GRASS

Area under permanent grass in 1938: 1,576,889
    acres (638,158ha)
Area under temporary grass in 1938: 1,459,552
    acres (590,673ha)

Although permanent grass and temporary grass
were separately recorded in 1938, as they were
in 1870, the boundary between them is somewhat
arbitrary and, owing to the abandonment of
these terms in 1959 in favour of a division into
grass seven years old and over and younger
grass, the figures given in this section are not
strictly comparable with those for 1965 and
1972. Grass was widely distributed throughout
the lowlands, with the western half of the central
lowlands, the Solway lowlands and the upper
part of the Tweed valley as the major areas
(Fig 266). The regional importance of the South
East is also confirmed by Table 84. However,
as Figs 267 and 268 and Table 84 show, temporary and permanent grass have two quite
different distributions: for whereas permanent
grass was most important throughout the south-
west and west, the principal area for temporary
grass was the north-east (though it should be

TABLE 84

Grass, Permanent Grass and Temporary Grass in 1938

High-land	North East	East Central	South East	South West	Scot-land
Percentage of grass in each region					
10	23	16	16	36	100
Percentage of permanent grass in each region					
12	9	15	17	47	100
Percentage of temporary grass in each region					
8	38	16	14	23	100
Percentage of grass under permanent grass					
61	20	51	57	68	52

Source: *Agricultural Statistics 1938*

noted that the same shadings on the two maps
represent different intervals); as on so many of
the agricultural maps of Scotland, the contrast
between Orkney and Shetland is also apparent.
In large measure, these differences reflect the
varying proportion of land under tillage, but
there can be little doubt that temporary grass
occupied a special place in the agriculture of the
north-east. It should, however, be noted that
there are considerable local differences, as the
parish statistics show, with permanent grass
replacing temporary grass at the upper margin
of cultivation.

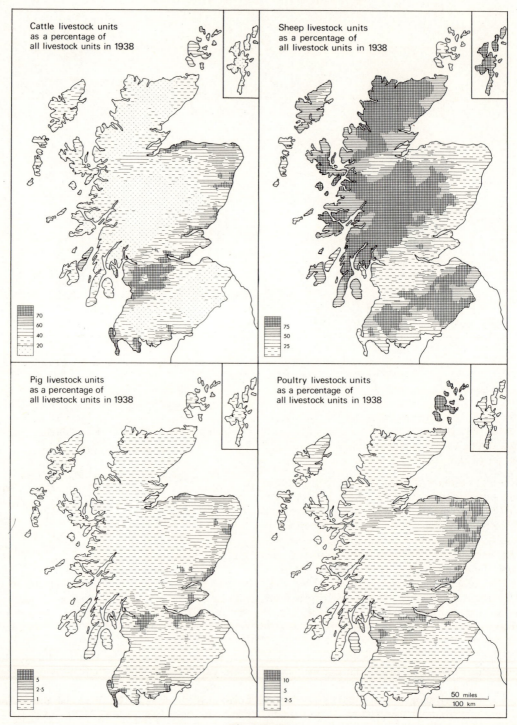

Cattle livestock units
as a percentage of
all livestock units in 1938

70
60
40
20

Sheep livestock units
as a percentage of
all livestock units in 1938

75
50
25

Pig livestock units
as a percentage of
all livestock units in 1938

5
2·5
1

Poultry livestock units
as a percentage of
all livestock units in 1938

10
5
2·5

50 miles
100 km

Figs 269–272

TABLE 85

Livestock Units in 1938

	Highland	North East	East Central	South East	South West	Scotland
			Number in thousands			
	382	433	304	325	715	2,158
		Percentage of livestock units, by type				
Beef cattle	8	30	26	14	8	16
Dairy cattle	18	19	15	10	42	25
Sheep	66	29	44	66	38	46
Pigs	1	3	3	3	2	2
Poultry	2	8	4	3	4	4
Horses	5	11	8	5	5	7
	100	100	100	100	100	100

Source: *Agricultural Statistics 1938*

LIVESTOCK UNITS

Number of livestock units in 1938: 2,158,239

Livestock units were computed from the 1938 census using similar factors to those employed in 1965, viz cows, 1; other cattle two years old and over, $\frac{3}{4}$; other cattle one year old and under two, $\frac{1}{2}$; other cattle under one year old, $\frac{1}{4}$; ewes and rams, $\frac{1}{5}$; other sheep, $\frac{1}{15}$; sows, $\frac{1}{2}$; boars, $\frac{2}{5}$; other pigs, $\frac{1}{7}$; poultry six months old and over, $\frac{1}{50}$; other poultry, $\frac{1}{200}$; horses, 1. While results are not strictly comparable with those for 1965, chiefly owing to the omission of horses, and while the choice of factors can be challenged, there is no doubt that stocking densities increased by between about a fifth and a quarter, despite the loss of agricultural land to urban expansion in the lowlands and to forestry in the uplands.

Table 85 shows the importance of the different kinds of livestock in 1938 in the five regions. The table also demonstrates the contrast between the absolute and relative importance of beef and dairy cattle, while comparison with Table 76 reveals the increasing importance of cattle; for, although the figures are not strictly comparable, the change is sufficiently large to justify this conclusion.

CATTLE LIVESTOCK UNITS

Number of cattle livestock units in 1938: 873,767

Cattle were the leading type of livestock in two principal areas in 1938, the lowlands of the south-west, especially the western half of the central lowlands, and the north-east (Fig 269). Of course, as Table 85 reveals, the proportions shown are somewhat misleading in that the relative importance of beef cattle and dairy cattle varies greatly, with dairy cattle far more important than beef in the South West and the Highlands and beef than dairy cattle in the remaining regions.

SHEEP LIVESTOCK UNITS

Number of sheep livestock units in 1938: 997,090

To a large extent, the distribution of sheep livestock units is the complement of that of cattle livestock units (Fig 270). Sheep were everywhere the leading livestock in the uplands; among the lowland areas, values of 50 per cent and over were found only in the Merse.

PIG LIVESTOCK UNITS

Number of pig livestock units in 1938: 48,219

Pigs accounted for a small proportion of livestock units everywhere. Those parishes where the proportion exceeded 5 per cent were located mainly in the central lowlands, particularly around Edinburgh and Glasgow (Fig 271).

POULTRY LIVESTOCK UNITS

Number of poultry livestock units in 1938: 94,350

Measured as livestock units, poultry were more important than pigs, though it must be remembered that Figs 271 and 272 show only those pigs and poultry recorded on agricultural holdings. The distribution of poultry livestock units is also more highly regionalised than that of pigs.

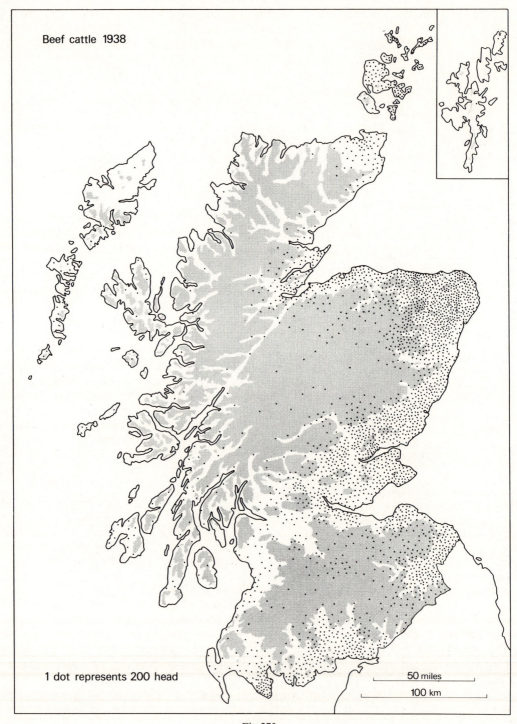

Beef cattle 1938

1 dot represents 200 head

50 miles

100 km

Fig 273

CATTLE
Number of cattle in 1938: 1,315,731

The number of cattle in Scotland was 26 per cent greater in 1938 than in 1870 (though some allowance must be made for improvements in the coverage of the agricultural returns), but only 65 per cent of that in 1965, an indication of the more rapid rise in the number of cattle in the intervening twenty-seven years. Of the cattle recorded in 1938, 56 per cent were returned by occupiers of holdings as dairy cattle.

BEEF CATTLE
Number of beef cattle in 1938: 581,317

Although beef cattle were less numerous than dairy cattle in 1938, they are given greater prominence in this section to assist comparison with maps of the situation in 1965 and also because the various stages of production have somewhat different distributions, a feature which is much less characteristic of dairy cattle. No comparison with 1870 is, of course, possible, since cattle were not then divided into beef and dairy cattle, but in 1938 the number of beef cattle was only 43 per cent of that in 1965; for the post-war period has been marked by a rapid expansion in beef production. Nevertheless, the distributions are broadly similar in both years, with most of the beef cattle being found in the eastern half of the country (Fig 273, cf Fig 135). The chief difference is that beef cattle were more widely distributed in 1965, though the North East was by far the most important region in both years (cf Tables 16 and 86). Fig 273 also shows the contrasts within the South West Region between the central lowlands and the lowlands of Dumfriesshire and Galloway; the Orkneys were the only island group with any considerable number of beef cattle.

Although the North East was the principal region, there were differences in the relative importance of the various categories of beef cattle; these are broadly summarised in Table 86 and are illustrated in more detail over the page in Figs 274–6.

Breeding cows were most numerous in the North East Region, but such cattle accounted for a higher proportion of the beef cattle in the Highlands, where 26 per cent of all beef cattle were beef cows, against a national average of 15 per cent. Such cows were less than a third as numerous as in 1965, when beef breeding was also more widespread and the importance of the North East and East Central Regions less marked (cf Tables 18 and 86). Other beef cattle two years old and over, were relatively more important in the East Central, South East and South West Regions, though the largest number were in the North East. This class of beef cattle was the only one which was less numerous in 1938 than in 1965, numbers falling by 24 per cent as a consequence of the trend towards slaughter of cattle at younger ages. Other beef cattle between one and two years old, on the other hand, more than doubled in number between 1938 and 1965, and the pre-eminence of the North East became more marked; in 1938, such cattle accounted for 35 per cent of the cattle in the North East, which had nearly half the national total. For both these categories of other beef cattle, the share of the Highlands was much smaller than the region's share of the breeding herd. Young beef cattle, ie those under one year old, were relatively most important in the Highlands, though, as numbers of older cattle show, many were subsequently exported to other regions; the North East, too, had above-average proportions of young cattle.

TABLE 86

Beef Cattle and Beef Cows in 1938

	Highland	North East	East Central	South East	South West	Scotland
Percentage of beef cattle in each region						
	9	40	22	13	16	100
Percentage of beef cows in each region						
	15	38	20	14	13	100
Percentage of other beef cattle two-years old and over in each region						
	4	33	27	15	20	100
Percentage of other beef cattle one-year old and under two in each region						
	8	44	21	11	15	100
Percentage of categories of beef cattle in each region						
Beef cows	27	15	14	16	12	15
Other beef cattle 2 years +	12	23	33	32	34	27
Other beef cattle 1–2	27	35	30	28	29	31
Other beef cattle −1	32	25	21	20	21	23

Source: *Agricultural Statistics 1938*

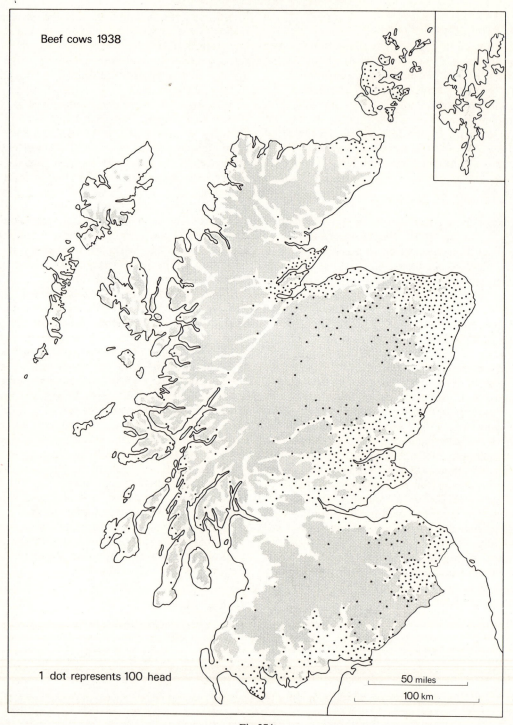

Beef cows 1938

1 dot represents 100 head

50 miles

100 km

Fig 274

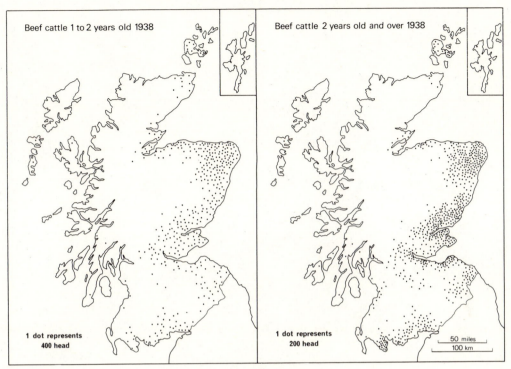

Figs 275–276

Beef Cows
Number of beef cows in 1938: 334,618

Beef cows were widely distributed throughout the lowlands of eastern Scotland, from the Merse to the Moray Firth, with an outlier in Caithness; they were also fairly numerous in the Solway lowlands, while only the Orkneys of the island groups had any large number (Fig 274). Not only were numbers much smaller than in 1965, but their distribution was less widespread and was largely a lowland one; the marked concentration along the margins of the uplands, which is such a distinctive feature of Fig 137, was notably absent, though the broad regional picture has changed relatively little.

Other Beef Cattle One year Old and Under Two
Number of other beef cattle one year old and under two in 1938: 181,394

Other beef cattle of this age group mainly represent store cattle, though some cattle being fattened for slaughter will also be included. Such cattle were mainly to be found in eastern Scotland, with major concentrations in the north-east (Fig 275); in 1965 they were not only much more numerous, but were more widely distributed in western Scotland. In comparing this map with Fig 276, showing the distribution of older cattle, it should be noted that one dot represents twice as many animals on the former map as on the latter.

Other Beef Cattle Two years Old and Over
Number of other beef cattle two years old and over: 158,767

Most of these older cattle were probably being fattened for slaughter; apart from the reduction in numbers, the distribution in 1965 is very similar to that in 1938, with the largest number in Aberdeenshire and adjacent counties (Fig 276). There has been a notable decline in numbers of such animals in the Solway lowlands and in Orkney.

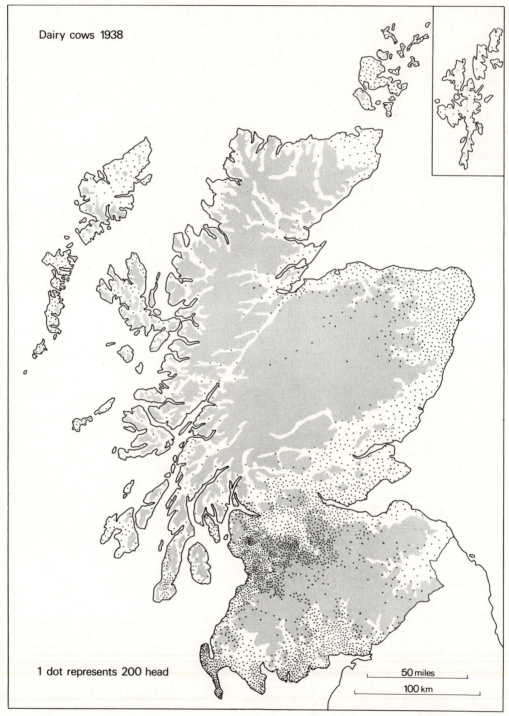

Dairy cows 1938

1 dot represents 200 head

50 miles

100 km

Fig 277

DAIRY CATTLE

Number of dairy cattle in 1938: 734,414

Dairy cattle and dairy cows (which accounted for 46 per cent of all dairy cattle in 1938) have very similar distributions and only that of dairy cows is mapped here (Fig 277). As in the 1960s, many farmers bred replacements for their own herds, but, as the ratio of dairy cows (ie cows and heifers in milk and cows in calf) to in-calf heifers shows, there was some rearing of heifers as herd replacements in other parts of the country; for the ratio ranged from approximately 13 to 1 in the Highlands to less than 4 to 1 in the South West, compared with a national average of 5 to 1 (Table 87). No direct comparison with dairying in 1870 is possible because dairy cattle were not then separately distinguished in the agricultural returns, but numbers in 1965 were rather smaller, at 94 per cent of the 1938 total, having reached a peak of 852,653 in 1950.

DAIRY COWS

Number of dairy cows in 1938: 334,618

In 1938 dairy cows were to be found mainly in the lowlands of west-central and south-west Scotland, especially in Ayrshire and in the Rhinns of Galloway and, to a lesser extent in the counties of Kirkcudbright and Dumfries (Fig 277). Numbers of dairy cows elsewhere were much smaller, though they were widespread and

TABLE 87

Dairy Cattle and Dairy Cows in 1938

High-land	North East	East Central	South East	South West	Scot-land
Percentage of dairy cattle in each region					
13	16	8	6	57	100
Percentage of dairy cows in each region					
15	15	9	6	56	100
Ratio of dairy cows to in-calf heifers					
13·1	13·0	7·4	5·3	3·6	5·0

Source: *Agricultural Statistics 1938*

were surprisingly numerous in the islands. The Lothians and the Merse had fewest dairy cattle, an interpretation confirmed by Table 87 which shows that the South East Region had only 6 per cent of all dairy cows in Scotland.

Until the twentieth century, the main dairying areas of the south-west were primarily concerned with the making of farmhouse cheese; supplies for milk for the urban markets, notably in the Clyde valley, were provided either from urban cowhouses (of which numbers still remained in the 1930s) or from nearby farms. Severe competition from imported cheese from the late nineteenth century onwards encouraged those making cheese to switch to the liquid market where this was possible, and milk was sent as far as London in the early 1930s when the price for liquid milk was being undercut by those switching from cheese making or from selling milk for manufacture; for milk was increasingly being processed in manufacturing creameries rather than on farms. The creation of the Scottish Milk Marketing Board (with responsibility for south Scotland) stabilised the market by introducing a common price to farmers irrespective of the use to which milk was put, so eliminating the tendency for distant producers to invade the liquid market. At the same time, proximity ceased to have any great value, partly because of improvements in road transport and a policy of tapering transport rates, so that a distant producer paid little more than a nearby one. Dairying spread to all those parts of the south-west which were suitable for dairying, with milk from distant producers still going mainly to manufacture except in winter when such areas often contributed to urban milk supplies. Elsewhere, local markets were largely served by dairy farms in close proximity.

Although numbers of dairy cows rose to a peak of 371,424 in 1950, the trend since has been towards a progressive concentration of dairying in the South West as farmers elsewhere gave up dairying, especially in eastern Scotland where dairying was often a subsidiary enterprise and other enterprises were possible (cf Tables 36 and 87).

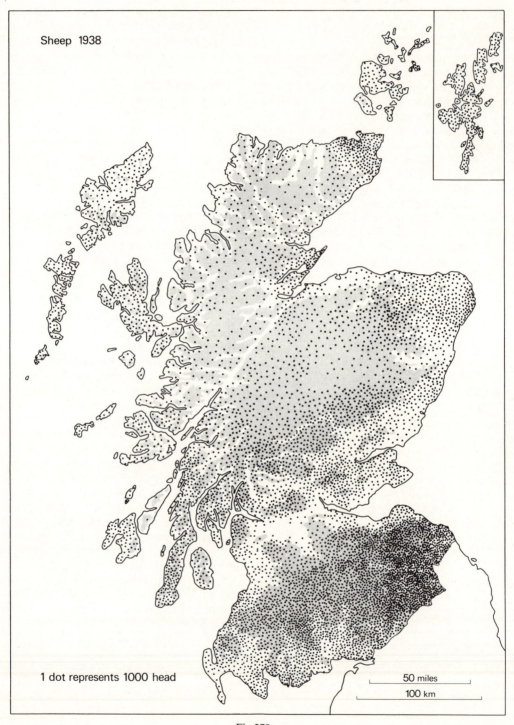

Sheep 1938

1 dot represents 1000 head

50 miles

100 km

Fig 278

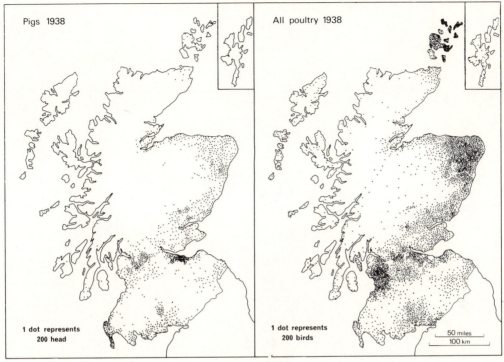

Figs 279–280

SHEEP

Number of sheep in 1938: 7,969,482

Sheep were 18 per cent more numerous in 1938 than in 1870, though there were only 93 per cent as many as in 1965. Fig 278 shows the continuing importance of the Southern Uplands and the Tweed valley; the main difference compared with 1870 is the lesser importance of the Highlands and the larger number of sheep in the North East, with 35 per cent of all sheep in 1870 (Table 88).

TABLE 88
Sheep, Pigs and Poultry in 1938

High-land	North East	East Central	South East	South West	Scot-land
Percentage of sheep in each region					
24	13	13	22	27	100
Percentage of pigs in each region					
5	23	17	18	37	100
Percentage of poultry in each region					
8	38	14	9	31	100

Source: *Agricultural Statistics 1938*

PIGS

Number of pigs in 1938: 257,374

The distribution of pigs was predominantly lowland and much more patchy, with notable concentrations around Edinburgh and Glasgow and in the Rhinns of Galloway (Fig 279). There were 62 per cent more pigs in 1938 than in 1870, but less than half as many as in 1965; the major difference was the decline in pig-keeping in the South West and in the Highlands, and the rise in importance of the North East (Table 88).

POULTRY

Number of poultry in 1938: 7,778,924

The three most important areas for poultry in 1938 were lowland Aberdeenshire, Orkney and the South West, especially lowland Ayrshire (Fig 280 and Table 88). The larger proportion of poultry in East Central and South East Scotland in 1965 reflects the changing character of a poultry industry some 12 per cent larger.

AN AGRICULTURAL ATLAS OF SCOTLAND

Agricultural trends 1870–1970

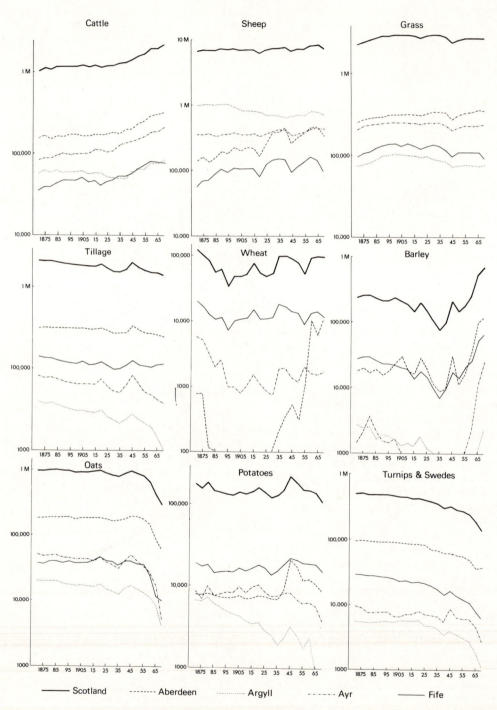

Cattle Sheep Grass

Tillage Wheat Barley

Oats Potatoes Turnips & Swedes

—— Scotland ------ Aberdeen .... Argyll ––·–· Ayr —— Fife

Crop figures are in acres

Fig 281

TRENDS 1870–1970

Figs 281, 282 and 283 sum up the changes in the period covered by this atlas, Fig 281 showing the broad trends over the hundred years since 1870 and Figs 282 and 283 recording, in more detail, changes since 1939. The scales are semi-logarithmic so that the slope on each line measures the rate of change, which is comparable between groups on the same scale. Figures for Scotland are shown by the thick line at the top of each graph and selected counties, viz Aberdeen, Argyll, Ayr and Fife, have been chosen to show the main regional trends. Of course, no county is ideal from this point of view; they vary greatly in size and are generally heterogeneous in physical character, often including both high and low ground and, in the case of Inverness and Ross and Cromarty, extending from coast to coast. Obviously, too, more counties could have been added to illustrate other regional trends, notably in the south-east (Berwickshire) and south-west (Wigtown). Of those chosen, Aberdeen represents the principal area for livestock (cattle) rearing and fattening, Argyll typifies the Highlands, Ayrshire the main dairying area of the central lowlands, while Fife represents the chief areas of crop production in eastern Scotland. They are contained within (and largely share the character of) the North East, Highland, South West and East Central Regions respectively. None is as compact or homogeneous as is desirable; nevertheless, they represent the best available units. The graphs should also be used to place the preceding cross-sections for 1870, 1938 and 1965 in perspective, and enable the reader to see these static distributions as a product of complex processes of change.

Because of the nature of the available data it is possible to take only a very limited view of livestock changes. Between 1870 and 1970 the number of cattle increased fairly steadily from 1,041,434 to 2,233,720, although the steepest rise came after 1930. A part of this early increase can probably be attributed to more complete enumeration (though the familiarity with form-filling which the Highland and Agricultural Society experiments had provided and the lower proportion of estimates probably made such a source of error less likely in Scotland than in England). Trends are similar in three counties, but in Argyll a slow decline was recorded between 1870 and the 1930s.

Changes in numbers of sheep were quite small, from 6,750,854 in 1870 to 7,493,866 in 1970, despite the loss of perhaps a million acres of agricultural land (mainly rough grazing) to afforestation in the interval. There are also greater year-to-year variations, a not surprising feature in view of the severe conditions under which many of Scotland's sheep population are kept. Trends in the four counties, however, diverge, with numbers tending to rise in the eastern counties of Aberdeen and Fife, to be stable in Ayr and to decline in Argyll.

The trends in land use, particularly the graphs of the acreage under grass, should also be treated with caution. The acreage under grass rose in the last third of the nineteenth century, partly through the conversion of land under tillage to grass and partly, no doubt, from improvements in the completeness of the returns. Thereafter the acreage changed little apart from small decreases during the ploughing campaigns of the two world wars. Here, too, there are differences between the county trends, with the acreage in Aberdeen increasing and those in Argyll and Fife declining (after a rise in the nineteenth century), though probably for different reasons; for, as the graphs of tillage show, the trend in Argyll reflects a decline in the total acreage of crops and grass and that in Fife a switch from grassland to tillage. The tillage acreage, on the other hand, shows a fairly steady decline (apart from the breaks during the two world wars), from a maximum of 2,145,723 acres (868,362ha) in 1870 to a minimum of 1,469,786 acres (594,814ha) in 1970. All the county graphs show a general downward trend, but the rate of change varied, being greatest in Argyll (though the acreage involved was small) and showing some tendency to reversal in Fife from the 1930s.

The patterns of cropping are more varied, though the method of representation exaggerates the importance of annual fluctuation among the minor crops where only small acreages are grown. The total acreage under wheat has changed little from the maximum of 125,642 acres (50,847ha) in 1870, though this statement conceals a decline in the late nineteenth century

Cropping changes 1939 – 1970

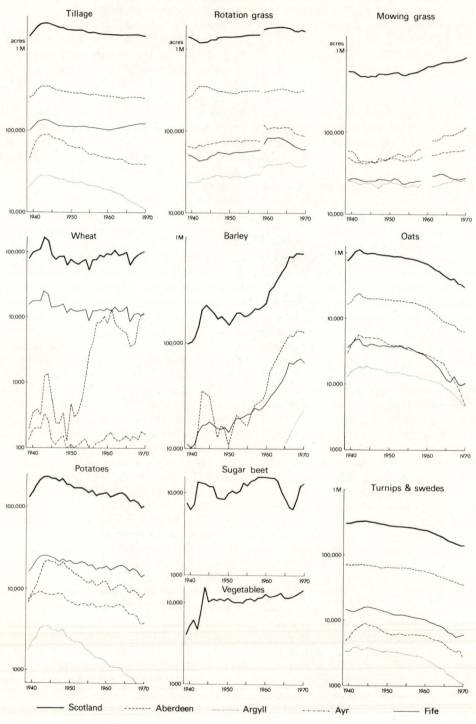

Fig 282

TABLE 89

Average Yields of the Principal Crops 1937–41 and 1968–72

	Wheat	Barley	Oats	Potatoes	Turnips and swedes
		cwt/acre			
Five-year average				tons/acre	
1937–41	22·4	19·5	16·3	7·3	17·1
1968–72	35·4	32·8	26·5	10·5	23·5
Annual					
1939	21·8	19·1	15·8	7·7	16·1
1970	36·8	32·7	25·6	11·4	25·8

Source: *Agricultural Statistics*

and a subsequent, though fluctuating, rise; the most important change was in Aberdeenshire, though the acreage grown in 1870 was only 755 acres (306ha). The acreage under barley fluctuated more markedly, declining from 244,142 acres (98,804ha) in 1870 to a minimum of 59,808 acres (24,204ha) in 1933, and rising steeply in the post-war period to a maximum of 718,524 acres (290,783ha) in 1969. All counties follow the same trend, though the rise in the two western counties occurred somewhat later. The area devoted to oats (1,019,596 acres, or 412,624ha, in 1870) remained fairly steadily until the 1940s, when it fell rapidly to 309,353 acres (125,193ha) in 1970; this trend was repeated in the counties, though in Argyll and, to a lesser extent, Ayr, there was a slow decline from 1870.

The acreage under potatoes declined slowly from 180,169 acres (72,913ha) in 1870 to 119,940 acres (48,539ha) in 1897, rose sharply during World War II and has since declined; this pattern is repeated in Fife and, even more markedly, in Aberdeen, but in Ayr and even more in Argyll the trend was downwards throughout. The trend in the acreage under turnips and swedes, on the other hand, resembles that under oats, with a slow decline from a maximum of 498,932 acres (201,915ha) in 1870 and a steepening rate of decline in the post-war period to a minimum of 140,472 acres (56,848ha) in 1970; for these crops, both the trends and the rates are similar in the four counties.

CHANGES 1939–1970

Fig 282 shows, for land uses and crops, the changes over the last thirty-five years covered by the previous graphs, but in greater detail, and underlines the relatively small changes during World War II. It also confirms the general upward trend in the acreage under rotation grass, but the discontinuity in 1959 marks a change in definition, when the term, 'temporary' or 'rotation' grass was abandoned and grass was merely divided into that under seven years old and older grass, the former being approximately equivalent to rotation grass as previously recorded. As the graph shows, this resulted in an increase in the acreage of 'rotation grass' and figures before and after this break are not strictly comparable. Since 1959, the acreages in both Ayrshire and Fife have trended downwards. By contrast, the acreages of mown grass (including both hay and silage, with a small amount of dried grass) show a steady upward trend, from 559,506 acres (226,429ha) in 1939 to 822,712 acres (332,947ha) in 1970; this tendency was most marked in Aberdeenshire and in Ayrshire.

In addition to the crops shown in Fig 281, the acreages under sugar beet and vegetables are also recorded in Fig 282, though only for Scotland as a whole; however, both crops were highly localised (Figs 75 and 87) and the national figures can be taken to indicate trends in these areas also. The acreage under sugar beet has fluctuated considerably, but fell sharply between 1947 and 1949; this drop was followed by some recovery, but the sugar beet factory at Cupar has now closed and production has ceased. The acreage under vegetables increased sharply during World War II, largely as a result of a great expansion of the acreage under turnips and swedes and under cabbages; after some decline in 1954, the acreage has trended steadily upwards.

It will of course, be appreciated that references to acreages alone can be misleading, for yields have improved markedly over this period, so that even a declining acreage may be accompanied by

Livestock changes 1939–1970

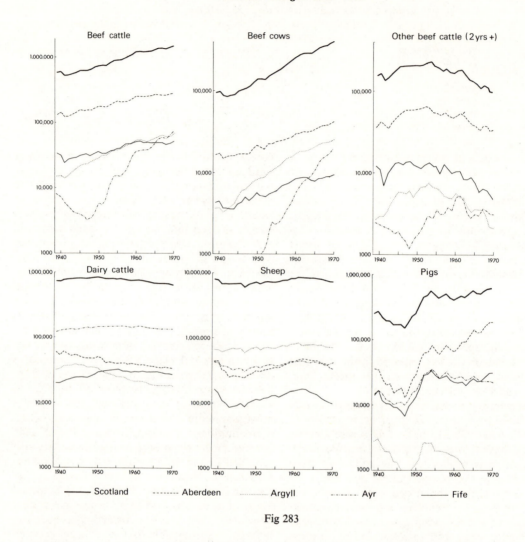

Fig 283

an increase in production. Table 89 shows, for the principal crops, the five-year average yield in 1937–41 and in 1968–72, and also the yields in 1939 and 1970.

Fig 283 shows in more detail for 1939–1970 the trends in numbers of livestock already discussed in relation to Fig 281. For this period, it is possible to analyse the differences between trends in numbers of beef and dairy cattle; whereas the former trend steadily upwards from 1941, when there were 536,276 beef cattle, to reach 1,567,616 in 1970, numbers of dairy cattle,

after reaching a peak of 852,653 in 1950, have fallen steadily to 666,104 in 1970.

All the sample counties follow the upward trend in numbers of beef cattle, though the wartime decline and subsequent recovery were proportionately greatest in Ayrshire. These upward trends were most marked in respect of breeding cows, which increased from 84,185 in 1942 to 412,518 in 1970, largely under the stimulus of the Hill Cattle Subsidy and the Calf Subsidy (and later the Beef Cow Subsidy) and because of better prices for fatstock. All counties followed the same trend, though it was more marked in the western counties of Argyll and Ayrshire, which both overtook Fife. Numbers of younger beef cattle also increased, those under one-year old rising from 120,284 in 1942 to 556,622 in 1970, and those between one-year old and two years from 168,552 to 427,407 in 1970. By contrast, numbers of older beef cattle two-years old and over, after rising from 137,662 in 1942 to a maximum of 233,747 in 1953, fell to a minimum of 99,144 in 1970, though Ayrshire was anomalous among the counties. Trends in dairy cattle, too, varied, for while numbers in Ayrshire and Fife had been maintained after the gains of the 1940s and 1950s, those in Aberdeen and Argyll fell steadily throughout the period. Yet, as with crop yields, there have also been steady improvements in milk yields of dairy cows during this period, rising in the Scottish Milk Marketing Board area from 573 gallons in 1938/9 to 825 gallons in 1970; as a result of such increases, the quantities of milk sold off Scottish farms rose by 86 per cent, despite the 4 per cent fall in the number of cows.

Graphs of numbers of sheep also highlight the downward trend in all areas in the 1960s, particularly in Fife. Numbers of pigs, after falling sharply during World War II, from 252,264 in 1939 to 167,600 in 1944, rose rapidly to 552,976 in 1954 and have since risen more slowly to 611,282 in 1970. The most notable feature of these graphs are the opposite trends in Aberdeen, illustrating the general rise in pig production in north-east Scotland, and the decline in Argyll (though numbers there have never been large). There were similar trends in numbers of poultry, which fell from 7,710,999 in 1939 to 6,147,248 in 1943, but have since risen to a total of 12,774,879 in 1970 (though this figure, nearly 3 million higher than in the preceding year, is not strictly comparable). Numbers of horses, which had fallen from 172,871 in 1870, to 141,561 in 1939, declined further and their collection was abandoned in 1961; an estimate in 1965 put the total at 2,385.

This examination of trends confirms the general stability of Scottish agriculture, especially in respect of its regional character. The most marked changes in both cropping and stocking have occurred since 1939 and in part reflect the increasing involvement of government in agriculture, particularly through the encouragement of beef cattle (though it could equally be argued that, without government support through the Hill Sheep Subsidy, the Winter Keep Scheme and price support for wool, numbers of sheep would have declined). Some changes reflect technical and scientific advances, such as the breeding of short-strawed, high-yielding barleys, which have progressively replaced oats as the leading cereal throughout most of the country. Nevertheless, the agriculture of Scotland in 1970 was recognisably, in its broad outlines, that in 1870, though with fewer farmers farming fewer acres. To what extent these regional identities will persist will depend in part on decisions taken in Brussels; yet it is unlikely that these can be so major that they can drastically recast an agriculture which is marked so strongly by the distinctive features of the Scottish environment.

Bibliography

THERE is a rich but uneven literature about the agriculture of Scotland, especially about those aspects which are relevant to an understanding of the maps in this atlas. As the items in this bibliography show, Scotland has a particularly rich source in the post-war period in *Scottish Agricultural Economics*. Other useful sources are *Scottish Agriculture* and the *Transactions of the Royal Highland and Agricultural Society* (now no longer published). The three Agricultural Colleges (North of Scotland, East of Scotland and West of Scotland) issue a steady stream of reports, dealing primarily with the profitability of enterprises in the region for which each college is responsible. General descriptive articles appear in *Scottish Agriculture* and in the handbooks prepared for the annual meetings of the British Association. The annual reports of the Department of Agriculture and Fisheries for Scotland provide valuable information, as do those of other statutory bodies such as the Crofters Commission and the Highlands and Islands Development Board; the Department also publishes an annual volume of agricultural statistics. Many of the books and articles listed below throw light on changes over time; the *Transactions of the Royal Highland and Agricultural Society* are very useful in this respect.

In addition to the books listed below, other accounts occur in studies of Great Britain or the United Kingdom as a whole; examples include:

Britton, D. K. *Cereals in the United Kingdom* (Oxford: Pergamon, 1969)

Committee on Fatstock and Carcase Meat Marketing and Distribution. *Report* (London: HMSO, 1969), Cmnd 2282

Ministry of Agriculture, Fisheries and Food. *Horticulture in Britain Part 1: Vegetables* (London: HMSO, 1967)

Ministry of Agriculture, Fisheries and Food. *Horticulture in Britain Part 2: Fruit and Flowers* (London: HMSO, 1970)

The list which follows represents only a selection of the available material. Priority has been given to those sources which deal with the whole of Scotland; the rich regional sources have largely been ignored.

INTRODUCTION

Coppock, J. T. *An Agricultural Atlas of England and Wales* (London: Faber & Faber, 2 Edin., 1976)

——. 'The Cartographic Representations of British Agricultural Statistics', *Geography*, 50 (1965), pp 101–14

——. 'An Agricultural Atlas of Scotland', *Cartographic Journal*, 20 (1969), pp 36–46

GENERAL

Advisory Panel on the Highlands and Islands. *Land Use in the Highlands and Islands* (Edinburgh: HMSO, 1964)

McVean, D. N. and Lockie, J. D. *Ecology and Land Use in Upland Scotland* (Edinburgh: Edinburgh University Press, 1969)

Scottish Council (Development and Industry). *Natural Resources in Scotland* (1961)

Select Committee on Scottish Affairs. *Land Resource Use in Scotland*, 5 vols, House of Commons Paper, 511 (London: HMSO, 1972)

Stamp, L. D. *The Land of Britain: Its Use and Misuse*, (London: Longmans, 1948)

Symons, J. A. *Scottish Farming Past and Present* (Edinburgh and London: Oliver and Boyd, 1959)

Tivy, J. (ed). *The Organic Resources of Scotland: Their Nature and Evaluation* (Edinburgh: Oliver and Boyd, 1973)

Urquhart, R. *Farming in Scotland* (London: D. Rendel, 1968)

Wood, J. *An Agricultural Atlas of Scotland* (London: Gill, 1931)

THE PHYSICAL BASIS OF AGRICULTURE

Bibby, J. S. 'Land Use Capability Assessment in Scotland', in J. Tivy (ed), *The Organic Resources of Scotland* (1973), pp 55–65

Birse, E. L. and Dry, F. T. *Assessment of Climatic Conditions in Scotland*, 1 (Macaulay Institute for Soil Research, Craigie Buckler, 1970)

Birse, E. L. and Robertson, L., ibid., 2 (1970)

Birse, E. L., ibid., 3 (1971)

Dunn, J. M. 'Farm Rents in Scotland', *Scottish Agricultural Economics*, 16 (1966), pp 27–31

Fitzpatrick, E. A. 'The Soils of Scotland', in J. H. Burnett (ed), *The Vegetation of Scotland* (Edinburgh: Oliver and Boyd, 1964), pp 36–62

Glentworth, R. 'Soils of Scotland', paper prepared for International Society of Soil Science meeting (September, 1966)

Gloyne, R. W. 'Notes and Some Speculations on the Effects of Climate on the Development of Horticulture in Scotland', *Scientific Horticulture*, 23 (1971), pp 22–40

——. 'The Climate of Scotland—a Brief Review', *Land Resource Use in Scotland*, Vol v (1972), pp 175–86

Green, F. H. W. 'Climate of the Scottish Uplands', *Advancement of Science*, 21 (1964), pp 4–8

Harper, P. C. 'Crop Growth and the Environment', *Scottish Agriculture*, 44 (1965), pp 172–6

Miller, R. 'Bio-Climatic Characteristics', ch 2 in J. Tivy (ed), *The Organic Resources of Scotland* (1973), pp 12–23

THE MAN-MADE FRAMEWORK OF FARMING

Beilby, O. J. 'Employment Trends in Scottish Agriculture', *Scottish Agriculture* 44 (1965), pp 180–3

Dunn, J. M. 'Workers Numbers 1948–63. Sources and Utilisation of Casual Labour', *Scottish Agricultural Economics*, 14 (1964), pp 262–6

——. 'The Age Structure of Scottish Farm Workers', *Scottish Agricultural Economics*, 18 (1968), pp 133–7

——. 'Some Features of Small Full-time and Large Part-time Farms', *Scottish Agricultural Economics*, 19 (1969), pp 205–20

Hendry, G. F. 'Scotland's Part-time Farms', *Scottish Agricultural Economics*, 12 (1962), pp 112–19

——. 'A Note on Farm Size in Scotland', *Scottish Agricultural Economics*, 13 (1963), pp 164–8

Hendry, G. F. and Beilby, O. J. 'The Small Farm in Scotland', *Scottish Agricultural Economics*, 8 (1958), pp 28–41

McEwan, L. V. 'The Progress of Mechanisation in Scottish Agriculture', *Scottish Agricultural Economics*, 7 (1957), pp 20–3

McIntosh, F. 'Changes in the Structure of the Scottish Farm Labour Force since 1951', *Scottish Agricultural Economics*, 17 (1967), pp 79–80

Millman, R. 'The Marches of the Highland Estates', *Scottish Geographical Magazine*, 85 (1969), pp 172–81

——. 'The Landed Properties of Northern Scotland', *Scottish Geographical Magazine*, 86 (1970), pp 186–203

——. 'The Landed Estates of Southern Scotland', *Scottish Geographical Magazine*, 88 (1972), pp 126–33

Moisley, H. A. 'The Highlands and Islands—a Crofting Region?', *Transactions of the Institute of British Geographers*, 31 (1962), pp 83–95

Scola, P. M. 'Scotland's Farms and Farmers', *Scottish Agricultural Economics*, 11 (1961), pp 59–62

Sparrow, T. D. 'The Mechanisation of Agricultural Production in Scotland', *Scottish Agricultural Economics*, 18 (1968), pp 140–5

Stewart, I. M. T. 'Farmers' Transport Costs in the Highlands', *Scottish Agricultural Economics*, 14 (1964), pp 223–45

Wagstaff, H. R. 'Scotland's Farm Occupiers', *Scottish Agricultural Economics*, 20 (1970), pp 277–85

LAND USE AND CROPS

Beilby, O. J. and Mackenzie, A. M. 'Potato Production in Scotland', *Scottish Agricultural Economics*, 13 (1963), pp 169–74

Department of Agriculture and Fisheries for Scotland. *Agricultural Survey of Scotland* (Edinburgh: HMSO, 1946)

Dickson, D. R. 'Changes in Land Use on Farms which Grow Sugar Beet', *Scottish Agricultural Economics*, 23 (1973), pp 217–21

Dunn, J. M. 'Scotland's Orchards', *Scottish Agricultural Economics*, 20 (1970), pp 286–9

——. 'The Location and Size Structure of Scottish Horticultural Production', *Scottish Agricultural Economics*, 21 (1971), pp 48–56

East of Scotland College of Agriculture. *Barley Production in Scotland*, Bulletin No 3 (Edinburgh, 1972)

Hay, F. G. *Report on the Marketing of Scotch Seed Potatoes*, Studies in Agricultural Marketing (University of Glasgow, 1969)

Holmes, J. C. 'Tillage Crops', ch 11 in J. Tivy (ed), *The Organic Resources of Scotland* (1973), pp 141–63

Hunt, I. V. 'The Grass Crop', ch 10 in J. Tivy (ed), *The Organic Resources of Scotland* (1973), pp 122–40

McEwan, L. V. 'Post-war Changes in Horticultural Cropping', *Scottish Agricultural Economics*, 10 (1960), pp 39–44

——. 'The Displacement of Oats by Barley in Scotland', *Scottish Agricultural Economics*, 14 (1964), pp 251–5

North of Scotland College of Agriculture. *The Economics of Oat Production in Scotland*, Economic Report, No 130 (Aberdeen, 1973)

Ross, J. A. 'Recent Trends in Production, Sales and Prices of Scottish Cereals', *Scottish Agricultural Economics*, 23 (1973), pp 207–16

Sparrow, T. D. 'Economic Aspects of the Scottish Potato Crop', *Scottish Agricultural Economics*, 18 (1968), pp 115–27

LIVESTOCK FARMING

Agricultural Adjustment Unit. *Hill Sheep Farming Today and Tomorrow*, Workshop Report No 13 (University of Newcastle upon Tyne, 1970)

Beilby, O. J. 'Beef Cow Numbers in Scotland',

Scottish Agricultural Economics, 24 (1974), pp 277–83

Carlyle, W. J. *Some Aspects of the Geography of Livestock Movement in Scotland*, unpublished PhD thesis (University of Edinburgh, 1970)

——. 'The Marketing and Movement of Scottish Hill Lambs', *Geography*, 57 (1972) pp 10–7

——. 'The Away Wintering of Ewe Hoggs from Scottish Hill Farms', *Scottish Geographical Magazine*, 88 (1972), pp 100–14

——. 'The Movement of Irish Cattle in Scotland', *Journal of Agricultural Economics*, 24 (1973), pp 331–50

Committee on Hill Sheep Farming in Scotland. *Report*, Cmnd 6494 (London: HMSO, 1944)

Dunn, J. M. 'The Reduction in the Number of Milk Producers', *Scottish Agricultural Economics*, 13 (1963), pp 248–50

——. 'Lambs and Hogg Production in Scotland', *Scottish Agricultural Economics*, 17 (1967), pp 68–74

——. 'The Reduction in the Number of Egg Producers', *Scottish Agricultural Economics*, 23 (1973), pp 222–5

Fraser, A. 'The Scottish Sheep Industry Today: a Critical Review', *Transactions of the Royal Highland and Agricultural Society*, 5th Series 56 (1954), pp 23–34

Hunter, R. F. 'Hill Sheep and Their Pasture', *Journal of Ecology*, 50 (1962), pp 651–80

McEwan, L. V. *The Marketing of Store Livestock in Scotland*, Studies in Agricultural Marketing (University of Glasgow, 1965)

McIntosh, F. 'Fat Cattle Production in Scotland 1963/4', *Scottish Agricultural Economics*, 15 (1965), pp 357–63

Mackenzie, A. M. 'An Estimate of Broiler Production', *Scottish Agricultural Economics*, 14 (1964), pp 258–9

McQueen, J. D. W. *Studies in the Scottish Dairy Industry*, unpublished PhD thesis (University of Glasgow, 1961)

——. 'Milk Surpluses in Scotland', *Scottish Geographical Magazine*, 77 (1961), pp 93–105

Milk Marketing Boards in Scotland. *The Changing Structure of Scottish Milk Production* (Glasgow, 1965)

——. *Milk Production '69* (Paisley, 1969)

——. *The Structure of Scottish Milk Production at 1972* (Paisley, 1972)

Robertson, J. C. 'Sheep and Cattle Systems of Hill Land', *Advancement of Science*, 21 (1964), pp 33–7

Robinson, J. F. *Survey of Blackface Sheep* (Edinburgh: HMSO, 1953)

Robson, N. 'Sources of Scottish Beef Supplies

1964–5', *Scottish Agricultural Economics*, 17 (1967), pp 55–6

——. 'Agricultural Broiler Production in Scotland', *Scottish Agricultural Economics*, 17 (1967), pp 75–7

Scola, P. M. 'Recent Trends in Egg Production', *Scottish Agricultural Economics*, 16 (1966), pp 24–6

Scottish Milk Marketing Board. *Scottish Dairy Farm Census 1964* (Glasgow, 1964)

Wright, A. 'Fat Cattle—Age at Slaughter', *Scottish Agricultural Economics*, 20 (1970), pp 273–6

——. 'Supplies and Requirements of Animal Feeding Stuffs in Scotland', *Scottish Agricultural Economics*, 20 (1970), pp. 292–6

AGRICULTURAL ENTERPRISES AND TYPES OF FARMS

Department of Agriculture and Fisheries for Scotland. *Types of Farming in Scotland* (Edinburgh: HMSO, 1952)

Dunn, J. M. 'Enterprise Specialisation in Eastern Scotland', *Scottish Agricultural Economics*, 21 (1971), pp 25–6

Mackenzie, A. M. 'Agricultural Output in Scotland by Regions 1961–2', *Scottish Agricultural Economics*, 15 (1965), pp 334–45

Ministry of Agriculture, Fisheries and Food *et al*. *The Changing Structure of Agriculture* (Edinburgh: HMSO, 1970)

Russell, T. J. 'The Size Structure of Scottish Agriculture', *Scottish Agricultural Economics*, 20 (1970), pp 299–325

Scola, P. M. 'An Economic Classification of Scottish Farms Based on the June Census 1962', *Scottish Agricultural Economics*, 15 (1965), pp 295–329

AN HISTORIC PERSPECTIVE

Beilby, O. J. 'Changes in Scottish Ewe Numbers', *Scottish Agricultural Economics*, 9 (1959), pp 29–36

Department of Agriculture and Fisheries for Scotland. *Agricultural Survey of Scotland* (Edinburgh: HMSO, 1946)

Ministry of Agriculture, Fisheries and Food. *A Century of Agricultural Statistics, Great Britain 1866–1966* (London: HMSO, 1968)

Scola, P. M. 'Changes in the Geographical Distribution of Cows', *Scottish Agricultural Economics*, 6 (1956), pp 24–6

Stamp, L. D. (ed.) 'County Reports of the Land Utilisation Survey, various authors, various dates in the 1940s

Whitby, H. 'Some Developments of Scottish Farming Since the War', *Journal of Agricultural Economics*, 21 (1970), pp 1–2

——. 'Beef Cow Numbers in Scotland', *Scottish Agricultural Economics*, 24 (1974), pp 277–83